WIRELESS TECHNOLOGY

Advances and Innovations for Smart IoT Applications

WIRELESS TECHNOLOGY

Advances and Innovations for
Smart IoT Applications

Rajesh Kumar Dwivedi, PhD
Vinod Kumar Singh, PhD
Raghav Dwivedi, PhD
Editors

APPLE ACADEMIC PRESS

First edition published 2026

Apple Academic Press Inc.
1265 Goldenrod Circle, NE,
Palm Bay, FL 32905 USA

760 Laurentian Drive, Unit 19,
Burlington, ON L7N 0A4, Canada

CRC Press
2385 NW Executive Center Drive,
Suite 320, Boca Raton FL 33431

4 Park Square, Milton Park,
Abingdon, Oxon, OX14 4RN, UK

Library and Archives Canada Cataloguing in Publication

CIP data on file with Canada Library and Archives

Library of Congress Cataloging-in-Publication Data

CIP data on file with US Library of Congress

ISBN: 978-1-998511-48-8 (hbk)
ISBN: 978-1-998511-49-5 (ebk)
DOI: 10.1201/9781998511495

Dedication

*Dedicated to the pioneers of wireless technology,
whose ingenuity fuels the advancements driving
smart IoT applications forward.*

About the Editors

Rajesh Kumar Dwivedi, PhD
Director, College Development Council at CSJM University, Kanpur, India

Rajesh Kumar Dwivedi, PhD, is currently working as a Director of the College Development Council at Chhatrapati Shahu Ji Maharaj University (CSJM) University, Kanpur, India. Early on, he was a collaborator with CeFSA, Trento, Italy, (1997) and was awarded a prestigious ICTP-TRIL fellowship (1998) at the University of Milan, Italy. He is a well-known teacher, researcher, and professor of physics with 36 years of teaching and research experience. His research areas of interest are laser plasmas, laser ablation deposition of thin films and their characterization, photo-fragmentation of supersonic fullerene beams, nanoparticles synthesis and antenna design for wireless communication, etc. He has published several research papers both in reputed national and international journals and has supervised many PhD, MPhil, and master's dissertations. He received his MTech in Materials Science and PhD in Physics from IIT, Kanpur in 1986 and 1998 respectively.

Vinod Kumar Singh, PhD
Professor and Dean (R&D), S.R. Group of Institutions (CEMT), Jhansi, UP, India

Vinod Kumar Singh, PhD, is currently working as a Professor and Dean (R&D) at the S.R. Group of Institutions (CEMT), Jhansi, UP, India. He has approximately 20 years of experience in electrical and electronics engineering. Prof. Singh is a senior member of the International Association of Computer Science and Information Technology (IACSIT) and the International Association for the Engineers and Computer Scientists (IAENG). He is also a member of the Institute of Electrical and Electronics Engineers (IEEE). He has worked as a coordinator for the National Program on Technology Enhanced Learning (NPTEL) at IIT Kanpur and as a nodal coordinator for Virtual Labs at the Indian Institute of Technology Roorkee.

Prof. Singh has published over 200 research papers in esteemed international journals, including journals published by IEEE, Springer, Elsevier, and Wiley. He is the author of six edited books published by renowned publishers such as Springer, Elsevier, and IGI Global (USA), as well as two reference books. His contributions include more than 20 SCI journals and 40 book chapters in Springer, along with five granted international patents. He has guided 10 PhD scholars and over 30 MTech students. He has received significant awards, including a best faculty award in research and development and best paper awards at various international conferences. He has also guided projects selected for financial grants under the Council of Science and Technology, Government of UP (CST UP). Prof. Singh has delivered expert lectures at numerous seminars and workshops and has organized various interactive workshops and seminars. He has served as an external examiner for PhD defense vivas at multiple universities and has chaired sessions at IEEE conferences, Springer conferences, and other international conferences. Additionally, he is a reviewer for several prestigious SCI journals and both international and national conferences.

He received his PhD in Electronics and Communication Engineering from B.U. Rajasthan in 2013, and his MTech in Digital Communication System from Bundelkhand Institute of Engineering and Technology, Jhansi, in 2009. He earned his BTech in Electrical Engineering from the Institute of Engineering and Technology, MJP Rohilkhand University, Bareilly, India.

Raghav Dwivedi, PhD

Assistant Professor, Department of Electronics and Communication Engineering, Pranveer Singh Institute of Engineering and Technology, Kanpur, Uttar Pradesh, India

Raghav Dwivedi, PhD, is currently working as an Assistant Professor in the Department of Electronics and Communication Engineering at the Pranveer Singh Institute of Technology, Kanpur, Uttar Pradesh, India. He completed his PhD in Electronics and Communication Engineering with a specialization in antenna design from the Bundelkhand Institute of Engineering and Technology, Jhansi, Uttar Pradesh,

India. Dr. Dwivedi has demonstrated his expertise through numerous publications and patents in his field. His research focuses on advancing the field of antenna design, VLSI, etc., aiming to contribute significantly to the realm of electronics and communication. His research endeavors not only reflect his intellectual acumen but also his unwavering determination to address pressing challenges in the field. Dr. Dwivedi's multidisciplinary approach and collaborative mindset have earned him recognition among peers and mentors alike, positioning him as a promising leader in the realm of advanced technology. With a strong academic background, he holds a master's degree in Embedded Systems.

Contents

Contributors

Saiyed Tazen Ali
Department of Electronics and Telecommunication Engineering, IET, Bundelkhand University, Jhansi, Uttar Pradesh, India

Zakir Ali
Department of Electronics and Telecommunication Engineering, IET, Bundelkhand University, Jhansi, Uttar Pradesh, India

Jobin K. Antony
Department of Electronics and Communication Engineering, Rajagiri School of Science and Technology, Kerala, India

Parul Awasthi
Department of Electronics and Communication Engineering, UIET, CSJM University, Kanpur, Uttar Pradesh, India

Vishal Awasthi
Department of Electronics and Communication Engineering, UIET, CSJM University, Kanpur, Uttar Pradesh, India

Vikram Bali
Department of Computer Science and Engineering, IMS Engineering College, Ghaziabad, Uttar Pradesh, India

Korhan Cengiz
Department of Computer Engineering, Istinye University, Istanbul, Türkiye

Sleebi K. Divakaran
Department of Electronics and Communication Engineering, Rajagiri School of Science and Technology, Kerala, India

Raghav Dwivedi
Assistant Professor, Department of Electronics and Communication Engineering, Pranveer Singh Institute of Engineering and Technology, Kanpur, Uttar Pradesh, India

Rajesh Kumar Dwivedi
School of Basic Sciences, CSJMU, Kanpur, Uttar Pradesh, India

Saurabh Dwivedi
Department of Computer Science and Engineering, IMS Engineering College, Ghaziabad, Uttar Pradesh, India

Manoj Kapil
Department of Computer Science, Houston Tillotson University, Texas, USA

Parminder Kaur
Chitkara University Institute of Engineering and Technology, Chitkara University, Punjab, India

Bhawna Khokher
Department of Electronics and Communication, Bhagwant University, Ajmer, Rajasthan, India

Deepti Das Krishna
Department of Electronics, CUSAT, Kerala, India

Ajay Kumar
Department of Electronics and Communication Engineering, IMS Engineering College, Ghaziabad, Uttar Pradesh, India

Akhilesh Kumar
Department of Electronics and Communication Engineering, National Institute of Technology, Silchar, Assam, India

Niraj Kumar
Department of Electronics and Communication, Chhatrapati Shahu Ji Maharaj University, Kanpur, Uttar Pradesh, India

Durbadal Mandal
Department of ECE, Computer Aided Electromagnetic Design Laboratory, National Institute of Technology (NIT), Durgapur, West Bengal, India

Prabina Pattanayak
Department of Electronics and Communication Engineering, National Institute of Technology, Silchar, Assam, India

Surjeet Raikwar
ECE Department, Bundelkhand Institute of Engineering and Technology, Jhansi, Uttar Pradesh, India

Arun Raj
Department of ECE, Computer Aided Electromagnetic Design Laboratory, National Institute of Technology (NIT), Durgapur, West Bengal, India

Manish Sharma
Chitkara University Institute of Engineering and Technology, Chitkara University, Punjab, India

Sanjeev Sharma
Department of Electronics and Communication, New Horizon College of Engineering, Bangalore, Karnataka, India

V. K. Sharma
Department of Electronics and Telecommunication Engineering, Bhagwant University, Ajmer, Rajasthan, India

V. K. Singh
Department of Electronics and Telecommunication Engineering, CSE Ambabai, Jhansi, Uttar Pradesh, India

Vinod Kumar Singh
Professor and Dean (R&D), S.R. Group of Institutions (CEMT), Jhansi, UP, India

D. K. Srivastava
ECE Department, Bundelkhand Institute of Engineering and Technology, Jhansi, Uttar Pradesh, India

Gaurav Verma
Department of Electronics and Communication Engineering, IET, Jhansi, Uttar Pradesh, India

Ramesh Kumar Verma
Department of Computer Science and Engineering, IMS Engineering College, Ghaziabad, Uttar Pradesh, India

Shashikant Verma
Department of Electronics and Communication Engineering, IET, Jhansi, Uttar Pradesh, India

Abbreviations

3GPP	3rd generation partnership project
5G	fifth generation
5GC	5G-core
ACO	ant colony optimization
AMF	access and mobility management function
AMPS	advanced mobile phone services
Arch	architecture
BDMA	beam division multiple access
BER	bit error rate
CAGR	compound annual growth rate
CDMA	code division multiple access
CDPD	cellular digital packet data
CNT	carbon nanotubes
CoDuch	dual cluster heads cooperation
CP	circular polarization
CREMA	centre for research in electromagnetics and antennas
CSES	C-shape etching slot
CSFD	collaborative sensor fault detection
CSO	cuckoo seo services
CSSR	complementary split ring resonator
CST	computer simulation technology
CUPS	control and user plane separation
D2D	device-to-device communication
dB	decibels
dBi	decibels-isotropic
DECT	digital enhanced cordless communication
DEIN	data and energy central system
DELTA	deep-learning-based cooperative architecture
DGS	defected ground structure
DL	deep learning
DP	dynamic programming
DPSO	discrete particle swarm optimization
DRL	deep reinforcement learning
DSERs	device-selective energy requests
EBG	electromagnetic bandgap
EH	energy harvesting

EHC	extended hierarchical clustering
EH-IRSP	energy harvesting intelligent relay selection protocol
EHN	energy harvesting node
EIRP	effective isotropic radiated power
EM	electromagnetic
eMBB	enhanced mobile broadband services
EPC	evolved packet core
ESOR	energy-saving pioneering directing
FA	fractal element antenna
FBMC	filter bank multicarrier
FBR	front-to-back ratio
FCC	federal communications commission
FIT	finite integration technique
FTAOA	fault-tolerant task allocation algorithms
FTCT	fault tolerance employing repair points in clusters, & failure clustering topology
GA	genetic algorithm
GSM	global system for mobile communications
ICNIRP	international commission on non-ionizing radiation protection
ICT	information and communication technology
IoT	internet of things
ITU	international telecommunication union
IUCP	improved uneven cluster protocol
LF	feed length
LTE	long-term evolution
ME	magneto-electric
MIC	microwave-integrated-circuit
MIMO	multiple-input-multiple-output
ML	machine learning
MLEM	maximum likelihood event localization
MMC	massive machine communication
MMICs	monolithic-MIC
MMT	millimeter-wave technology
mMTC	massive machine type communication
MNs	moving networks
MSAs	microstrip patch antennas
NMT	nordisk mobile telephony
NOMA	non-orthogonal multiple access
NP	nanoparticle
NWCFs	nonwoven conductive fabrics
OFDM	orthogonal-frequency-division-multiplexing
OFDMA	orthogonal frequency division multiple access

PANI	polyaniline
PCB	printed circuit board
PCE	power conversion efficiency
PDMS	polydimethylsiloxane
PDR	packet delivery ratio
PMU	power management unit
PPy	polypyrole
PSO	particle swarm optimization
QoS	quality of service (QoS)
RF	radio frequency
RL	return loss
RSSI	received signal strength indicator
SAR	specific absorption rate
SBDTM	static fallback & dynamic timing tracking
SISO	single-input-single-output
SIW	substrate-integrated waveguide
SMF	session management function
SNR	signal-to-noise
SPA	simple patch antenna
SVM	support vector machine
TDMA	time division multiple access
TEMSEP	threshold-oriented and energy-harvesting enabled multi-level stable election protocol
TLBO	teacher learner based optimization
TRP	total radiated power
TSCH	time-slotted channel hopping
UAV	unmanned aerial vehicles
UCB	upper credibility bound
UDNs	ultra-dense networks
UMTS	universal mobile telecommunications system
UPF	user plane function
URC	ultra-reliable communication
uRLLC	ultra-reliable low latency communication
UWB	ultra-wideband
VNAs	vector network analyzers
VNI	visual network index
VSWR	voltage standing wave ratio
WBAN	wireless body area networks
WEH	wireless energy harvesting
WF	feed width
WiMAX	worldwide interoperability for microwave access
WSN	wireless sensor node

Preface

In an era defined by rapid technological advancements, the proliferation of wireless technology stands as a beacon of innovation, revolutionizing how we interact with the world around us. From smart homes to connected cities, from wearable devices to industrial automation, the transformative potential of wireless technology knows no bounds. It is in this dynamic landscape that the book *Wireless Technology: Advances and Innovations for Smart IoT Applications* emerges as a beacon of knowledge, offering invaluable insights into the latest trends, developments, and breakthroughs shaping the future of wireless communication and IoT ecosystems.

As the editors of this comprehensive volume, we are thrilled to present a collection of cutting-edge research, thought-provoking analyses, and practical applications that illuminate the multifaceted nature of wireless technology and its transformative impact on society. Through the contributions of esteemed experts and researchers from around the globe, this book delves into a diverse array of topics, spanning the realms of antenna design, signal processing, network protocols, security, and beyond.

Drawing upon the collective expertise of our contributors, this book explores the intricacies of wireless communication systems, shedding light on emerging technologies such as 5G, mmWave, and IoT, and examining their implications for industries ranging from healthcare and transportation to agriculture and manufacturing. Moreover, it addresses the challenges and opportunities inherent in the deployment of wireless networks, tackling issues of scalability, interoperability, privacy, and security in an increasingly interconnected world.

At its core, this book serves as a comprehensive resource for researchers, engineers, practitioners, and policymakers seeking to navigate the complex landscape of wireless technology and harness its transformative potential for the betterment of society. Whether you are a seasoned professional looking to stay abreast of the latest advancements or a curious newcomer eager to explore the frontiers of wireless innovation, this book offers something for everyone.

As editors, it has been our privilege to curate this collection of insights and perspectives, and we extend our heartfelt gratitude to all the contributors who have generously shared their expertise and knowledge. We hope that this book serves as a catalyst for dialogue, inspiration, and collaboration, propelling us toward a future where wireless technology empowers us to build smarter, more connected, and more sustainable communities. Together, let us embark on a journey of discovery and innovation as we unlock the boundless potential of wireless technology to shape a brighter tomorrow.

—*Editors*

Acknowledgments

In the pursuit of knowledge and the dissemination of wisdom, we stand on the shoulders of giants, guided by the collective wisdom, expertise, and generosity of those who have supported us along the way. As we embark upon the journey of publishing this book, *Wireless Technology: Advances and Innovations for Smart IoT Applications,* we are humbled and deeply grateful for the contributions of numerous individuals whose unwavering support and invaluable insights have made this endeavor possible.

First and foremost, we extend our heartfelt gratitude to the esteemed contributors whose scholarly contributions have enriched the pages of this book with their depth of knowledge, innovative ideas, and meticulous research. Their dedication to advancing the field of wireless technology and IoT applications is evident in the quality and rigor of their work, and we are honored to have had the opportunity to collaborate with such esteemed scholars and researchers. We are indebted to the reviewers whose constructive feedback and insightful critiques have helped refine and strengthen the content of this book, ensuring its relevance, accuracy, and scholarly integrity. Their commitment to excellence and their willingness to lend their expertise to the peer-review process are integral to the quality assurance of academic publications, and we are deeply grateful for their contributions.

We would also like to express our appreciation to the editorial and production teams whose professionalism, expertise, and dedication have played a vital role in bringing this book to fruition. From manuscript preparation to final publication, their tireless efforts and attention to detail have ensured the smooth progress and successful completion of this project.

Furthermore, we extend our gratitude to our colleagues, mentors, and academic institutions whose support and encouragement have been instrumental in our academic and professional development. Their guidance, mentorship, and encouragement have inspired us to pursue excellence and strive for impact in our respective fields of endeavor.

Last but certainly not least, we offer our deepest thanks to our families, friends, and loved ones whose unwavering support, understanding, and patience have sustained us throughout the course of this project. Their love, encouragement, and belief in our abilities have been a constant source of strength and inspiration, reminding us of the importance of balance, perspective, and gratitude in the pursuit of academic and professional success. We acknowledge with gratitude the myriad individuals whose contributions, support, and encouragement have enriched our lives and made this book possible. It is our sincere hope that the knowledge and insights contained within these pages will inspire future generations of scholars, researchers, and practitioners to continue pushing the boundaries of knowledge and innovation in the dynamic field of wireless technology and IoT applications.

—Editors

Introduction

In the intricate tapestry of modern technological discourse, the realm of wireless communication stands as a bastion of innovation, continuously evolving and reshaping the fabric of human interaction and connectivity. As we traverse the labyrinthine corridors of the digital age, the advent of wireless technology has heralded a paradigm shift in how we perceive, engage with, and navigate the world around us. It is within this dynamic milieu that we embark upon a journey of exploration and inquiry, delving into the multifaceted landscape of wireless technology and its profound implications for the emergent era of the Internet of Things (IoT).

At the intersection of connectivity and innovation lies the cornerstone of our discourse: the convergence of wireless technology and IoT applications. This convergence represents a nexus of opportunity and challenge, where the boundaries of possibility are pushed ever further by the relentless march of technological progress. In the crucible of this convergence, we find ourselves confronted with a myriad of questions, complexities, and opportunities, each demanding our attention and insight as we navigate the intricate interplay between theory and practice, vision and execution. As we embark upon this intellectual odyssey, our journey takes us through a landscape marked by towering peaks of technological achievement and sprawling valleys of uncertainty. We are guided by the steady beacon of inquiry, driven by a relentless pursuit of knowledge and understanding. Our quest is not merely to catalog the advancements of the past or prognosticate the trends of the future, but to engage in a deeper exploration of the underlying principles, mechanisms, and dynamics that govern the intricate dance of wireless communication and IoT integration.

In this introduction, we set the stage for our discourse, laying bare the foundational concepts, theoretical frameworks, and practical considerations that underpin the complex interplay between wireless technology and IoT applications. We delve into the historical antecedents of wireless communication, tracing its evolution from humble beginnings to the dizzying heights of technological sophistication we witness today. We explore the theoretical underpinnings of wireless communication systems, unraveling the mysteries of electromagnetic waves, modulation techniques, and signal processing algorithms that form the backbone of modern wireless networks. Moreover, we confront the myriad challenges and opportunities that lie on the horizon, from the quest for spectrum efficiency and network reliability to the imperative of privacy protection and security assurance in an era of ubiquitous connectivity. We examine the transformative potential of IoT applications across diverse domains, from healthcare and transportation to agriculture and smart cities, illuminating the myriad ways in which wireless technology is reshaping the fabric of society and redefining the boundaries of human experience. As we embark upon this intellectual odyssey, we invite the reader to join us on a journey of discovery and exploration, as we navigate the turbulent waters of technological innovation and chart a course towards a future defined by connectivity, creativity, and collaboration. Together, let us embark upon a voyage of intellectual discovery, as we unravel the mysteries of wireless technology and unlock the boundless potential of the Internet of Things.

Welcome to the brave new world of wireless technology and IoT applications. The journey begins here.

In the realm of wireless technology and IoT applications, this discourse navigates through various chapters:

Chapter 1 explores principles, technologies, and applications of RF energy harvesting systems for powering IoT devices wirelessly.

Chapter 2 investigates the design and analysis of fractal array antennas for 5G millimeter-wave bands, focusing on enhanced performance characteristics.

Chapter 3 covers design aspects of wearable antennas integrated into biomedical monitoring systems, addressing miniaturization, biocompatibility, and performance optimization.

Chapter 4 presents a comprehensive overview of a hybrid energy harvesting model inspired by biological mechanisms, combining multiple energy harvesting techniques for enhanced efficiency and adaptability.

Chapter 5 explores the design and optimization of microstrip patch antennas with notch and slot loading techniques for IoT/ISM/WLAN applications in the S-band frequency range.

Chapter 6 develops a novel slotted hammer type stretchable antenna tailored for IoT devices, addressing flexibility, durability, and compact form factor requirements.

Chapter 7 examines the symbiotic relationship between 5G communication technology and IoT ecosystems, discussing enhanced capabilities, applications, and emerging trends.

Chapter 8 mitigates interference in MIMO systems through Genetic Algorithm-based optimization techniques to enhance spectral efficiency and reliability.

Chapter 9 investigates design methodologies and practical implementations of mm-Wave MIMO antenna arrays to enable ultra-fast and reliable wireless communication in 5G networks.

Chapter 10 explores the development and optimization of multiband antennas on flexible substrates for wireless communication and sub-5G applications.

Chapter 11 details the design and fabrication of a square-shaped slot-cut microstrip patch antenna tailored for IoT applications, focusing on performance enhancement techniques. Each chapter provides in-depth insights and methodologies relevant to its specific topic, offering valuable resources for researchers, engineers, and practitioners in the field.

CHAPTER 1

RF Energy Harvesting Systems

SLEEBI K. DIVAKARAN,[1,*] DEEPTI DAS KRISHNA,[2] and JOBIN K. ANTONY[1]

[1]*Department of Electronics and Communication Engineering, Rajagiri School of Science and Technology, Kerala, India*

[2]*Department of Electronics, CUSAT, Kerala, India*

*Corresponding author

ABSTRACT

Energy harvesting is gaining attention for its ability to capture environmental energy and transform it into usable power, extending battery life. This thesis explores RF energy harvesting methods for powering low-energy IoT devices, wireless sensor nodes, and RFID, focusing on GSM and Wi-Fi bands. With the rise of compact low-power devices, sustainable powering methods are crucial. The research investigates rectennas, which receive RF energy and convert it for storage. The thesis comprises four sections, beginning with RF mapping to assess available ambient RF power. The second part explores multiport antenna designs that combine power from each antenna to enhance efficiency. Dual-band antennas mounted on six faces of a cube enable omnidirectional power reception, improving PCE. The rectenna uses a cost-effective FR4 substrate. The study develops single-band, dual-band, and wideband rectifiers to boost efficiency. Impedance matching circuits with radial stubs provide broad bandwidth matching. The Greinacher rectifier circuit achieves the desired DC power output for low incident power, though increasing stages improves voltage at the cost of PCE. The wideband rectifier maintains consistent efficiency across frequency and input power variations. The research advances wireless energy harvesting through innovative wideband rectifier designs and multiport antennas.

Wireless Technology: Advances and Innovations for Smart IoT Applications. Rajesh Kumar Dwivedi, Vinod Kumar Singh, and Raghav Dwivedi (Eds.)

DOI: 10.1201/9781998511495-1

1.1 INTRODUCTION

Recent advancements in technology have led to the development of the Internet of Things (IoT), 5G wireless systems, wearable electronics, and so forth, which require the extensive deployment of sensors with wireless communication capabilities that are positioned in faraway locations [1]. Since we must rely on battery sources, power is the limiting factor. This results in the laborious process of replacing and disposing of the massive quantity of batteries that pollute the environment. The issue can be solved environmentally or sustainably by using energy harvesting (EH), which entails gathering ambient energy and converting it into electricity to charge the batteries and sensors. There are many different types of ambient energy sources, including electromagnetic, thermal, mechanical, wind, solar, and tidal energy [2].

In order to transform incoming ambient energy into electrical energy that may be used in any application, a Power Management Unit (PMU) processes the electrical energy. The most plentiful source of energy, solar power, has a sporadic supply [3]. Radio frequency (RF) is a good substitute because it is always available, although its incident power levels are modest, which can be enhanced by employing an effective RF to DC conversion circuit and a dedicated source. This is a practical choice for energy harvesting because of the low power consumption of the sensors and the widely available RF power.

Figure 1.1 shows the RF energy harvesting system's block diagram along with the variables that impact the system's performance.

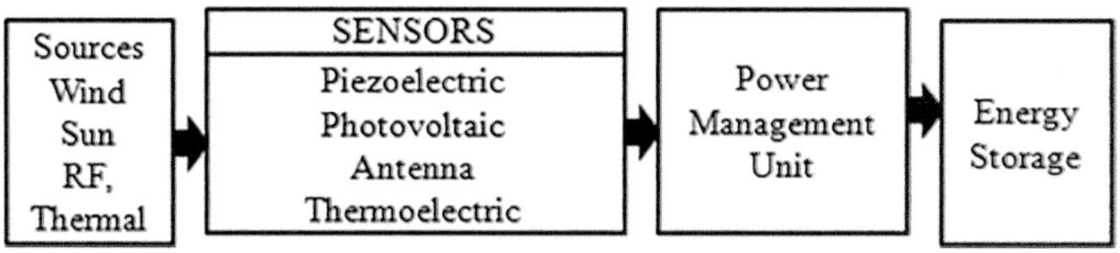

FIGURE 1.1 Block diagram of radio frequency (RF) energy harvesting system.

Source: Adapted from Ref. [4]

Additionally, the efficiency of RF energy harvesting has significantly improved in recent years, making it an attractive option for various applications. Moreover, the scalability and versatility of RF energy harvesting systems make them suitable for integration into a wide range of devices and environments. Furthermore, advancements in antenna design and signal processing techniques have further enhanced the performance and reliability of RF energy harvesting systems, making them more dependable for long-term use. Batteries are the main source of power for low-power devices and embedded systems. As they have a short lifespan, periodic replacement becomes necessary, and disposing of them in the environment causes pollution. This can be overcome with the help of self-sustainable and green energy harvesting techniques, such as harvesting ambient energy from the sun, wind, tidal, and electromagnetic sources [4].

Table 1.1 tabulates the different energy sources, corresponding power density, and harvesting techniques. Despite being the most plentiful source of energy, solar energy has an inconsistent supply and reach. Only 20% of the power delivered by the RF base station is used for communication and data transfer, and it is available continuously [5]. This makes RF energy harvesting systems a suitable choice for powering wireless sensor nodes, IoT devices, and so forth.

TABLE 1.1 Power Available From Various Sources [5]

Origin	Harvesting Mechanism	Power Density
RF	Rectenna	Wi-Fi – 1mW/cm^2 GSM – 0.1 μW/cm^2
Thermal	Thermoelectric	Industrial – 10 mW/cm^2
		Human – 30 μW/cm^2
Solar	Photovoltaic	Outdoor – 10 mW/cm^2
		Indoor: 10 μW/cm^2

Electrical energy for the RFEH process may come from an unidentified, ambient, or designated RF source. The range of power density available from the ambient environment is 0.1 μW/cm^2, and from a dedicated source, it is 1000 μW/cm^2. The RF power received decreases by a factor of the square of the distance and is dependent on frequency and distance according to Friis's law. Additionally, the Electro Technical Commission has set the standards and frequency for the safe functioning of electromagnetic devices [4]. The overall RF exposure restrictions specified by the FCC limit the incident RF power density that is available for RFEH. Dedicated sources and effective RF-to-DC conversion circuits can help reduce the limitation of low incident power levels. This is a practical choice for energy harvesting because of the low power consumption of the sensors and the widely available RF power.

The different sources of radio frequency energy include Wi-Fi, WLAN, digital TV, AM, FM, Bluetooth, and so on. It can power wireless sensor nodes [6], wireless body area networks [7], wireless charging systems [8], and RFID tags [9] because of its widespread usage.

The RFEH may be far-field or near-field based. Either inductive coupling [10] or magnetic resonance coupling [11] can be used to provide near-field RFEH. In the former, energy is exchanged between two systems by evanescent wave coupling, whereas in the latter, energy transfer is achieved through magnetic coupling between two coils that are resonating at the same frequency. The salient characteristics of this kind of system are incident power density and conversion efficiency. The power conversion efficiency (PCE) is, however, determined by the coupling coefficient, which depends on the distance between the coils [12]. But power transfer is restricted by distance and necessitates accurate coil alignment at both the transmitter and the receiver.

Far-field energy transfer is defined as the transfer of RF power at a distance greater than $\lambda/2\pi$ [13]. This makes it possible for RF energy to power devices dispersed across a large region. However, the distance between the transmitter and receiver limits the incident RF energy. The primary issue with this system is its low conversion efficiency [14]. The energy transmission properties of near-field and far-field are compared in Table 1.2. The near-field transfer is only appropriate for short distances and is used to power and charge wearable technology and cell phones. Far-field has the advantage of distance but still poses many research challenges, especially in improving its conversion efficiency [15].

TABLE 1.2 Radio Frequency (RF) Transmission Characteristics [16]

Characteristics	Inductive Coupling (Near Field Transfer)	Far Field Transfer	Resonant Coupling (Near Field Transfer)
Field	Magnetic method	Electromagnetic	Resonance
Distance	Short	Short to Long	Medium
Method	Coil	Antenna	Resonator
Regulation	Under discussion	Radio wave	Under discussion
Incident power	High	Low to High	High
Efficiency	High	Low to high depending on incident power	High

The primary design difficulty with the rectenna is its PCE, which quantifies how well a rectifier converts incoming RF energy into DC. The rectenna's RF to DC conversion efficiency is determined by Equation 1.1.

$$\eta_{RF-DC} = \frac{P_{DC}}{P_{RF}} \tag{1.1}$$

where P_{DC} and P_{RF} are the DC output power and the input RF power, respectively. Due to impedance mismatch, the rectifier's input voltage fluctuates depending on the frequency and intensity of the incident wave, which alters the diode's impedance and lowers the PCE. The following are some of the variables influencing the total power conversion efficiency of RF-EH systems:

- Unpredictable fading-related incident RF power availability.
- Lost components and restricted sensitivity.
- Low incident power as a result of the transmitter and receiver's separation distance and the maximum radiated power limitations.
- A strong nonlinearity exists between the output voltage and the input, particularly at low input powers.
- Trade-off between efficiency and bandwidth as a result of shrinking.

1.2 HISTORY OF WIRELESS ENERGY HARVESTING

The electrical energy is transmitted wirelessly by Nikola Tesla for the first time in an electric lighting system [17]. His work regarding wireless power transfer did not receive good acceptance during his lifetime. Tesla found in 1892 that certain high-frequency currents could be conveyed through low-pressure gases with ease. His curiosity about the prospect of powering electric gadgets located far from the energy source was piqued by this. He set up a large coil in his new laboratory in 1899 to generate 10 million volts and successfully powered several electric bulbs farther away from the source, which is the first demonstrated wireless power transfer. Guglielmo Marconi powered light bulbs wirelessly using RF in the 1920s. Heinrich Hertz published a paper describing wireless power transmission in 1941 [18].

Another development was made in 1964 by W. C. Brown, who powered a helicopter model using microwaves transmitted from the ground with the help of a rectenna. In the 1970s, the rectenna laid the foundation for powering solar-powered satellites, which can harvest energy from sunlight with the help of solar cells in space and beam it down to rectenna arrays placed on Earth using microwaves. These are the inventions that laid the foundation for wireless energy harvesting.

1.3 APPLICATION OF RF ENERGY HARVESTING

The first rechargeable lithium-ion battery was developed by combining a lithium cobalt oxide cathode and a carbon anode [19]. A battery, as a portable energy source, is limited by its lifetime and the presence of toxic materials, as well as requiring periodic recharging and replacements, which leads to environmental pollution.

RF energy harvesting is an emerging area that helps wireless nodes to work continuously, as this method helps to recharge them intermittently. Low-energy devices can be powered using this technology. A successful experimental setup of this technology is used for RFID applications [20]. These are examples of dedicated energy harvesting.

However, a variety of ambient radio frequency (RF) emissions can now be found in urban areas, thanks to the development of current communications technologies. Furthermore, there are now more prospects for new fields of use for ambient radio frequency energy harvesting due to the lower power requirements of sensor nodes utilized in autonomous systems. Several uses for this kind of energy include:

1. **Health Monitoring:** Power supply for devices like prostheses and cardiac pacemakers that need high uptime and for which installing batteries or any other traditional power source would be dangerous.

2. **Smart Sensors:** These include environmental monitoring systems that identify forest fires in situations where it is challenging to get close to the powered equipment for maintenance or when there isn't a connected power supply available.

3. **Embedded Sensors:** Such as those found, for example, beneath the asphalt of bridges and other structures where high uptime is required and it is not viable to attach batteries or other traditional power sources.

The nodes' power requirements determine the RFEH system's design specifications. Figure 1.2 displays the power consumption of different nodes. The area in Figure 1.2 where RFEH is currently applicable is shown by the highlighted circle. This range can frequently be expanded based on the type of energy source. Additionally, it displays small electronic devices like mobile phones and WSN nodes, as well as the types of batteries employed. As a result, one of the best options for alternative power for autonomous devices is the use of RF harvesting systems, which will also attract attention for future research since they can replace batteries and provide independent energy to lower maintenance and operating costs. Its reliance on such low RF power density incident levels, however, indicates that further development of this technology is required. Because it affects the overall efficiency of the autonomous system, choosing the operating frequency band and rectenna type for the RFEH system, depending on the intended application, may be essential.

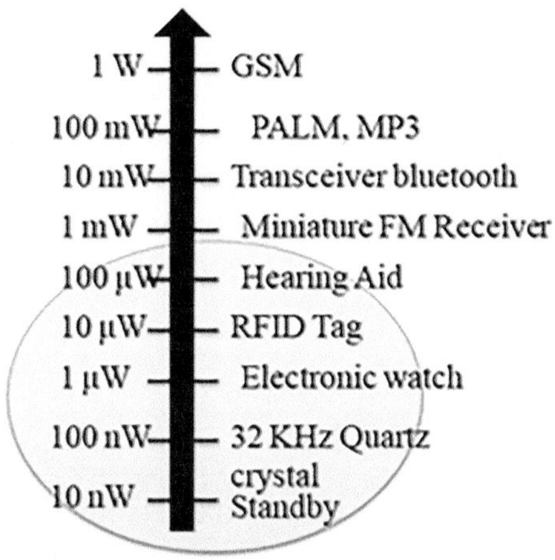

FIGURE 1.2 Power requirements of small electronics products.

1.4 RECTENNA

The rectenna is the core of the RFEH system. It is frequently utilized for WPT since it can convert RF energy into DC [29]. As seen in Figure 1.3, the antenna, harmonic suppression filter, rectifier, voltage multiplier, matching circuit, and load are the essential parts [30]. The matching circuit is used to match the antenna impedance with the remaining circuit in order to achieve high efficiency after the antenna receives the ambient or dedicated RF signals. In addition, it serves as a filter to keep out reflections [31]. The output voltage of the multiplier circuit is determined by the number of steps, and the rectifier is used to convert the RF impulses into DC [32]. The output DC is increased by a voltage multiplier because the incident RF amplitude is low.

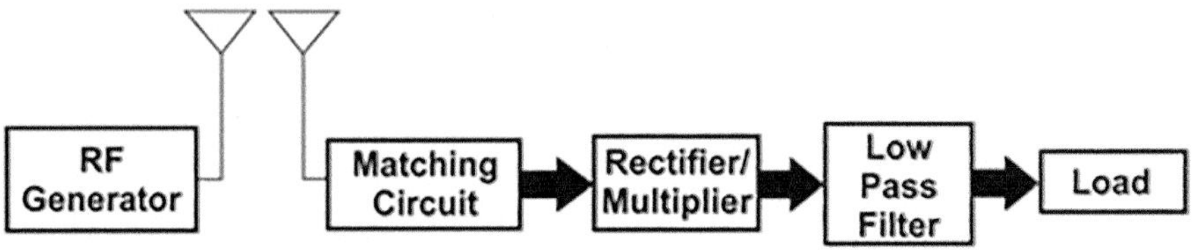

FIGURE 1.3 RF-EH system block diagram.
Source: Adapted from Ref. [4]

The main factor influencing the RFEH system's PCE is the rectenna. To achieve a high PCE, there are two methods. The highest power absorption and its transfer to the rectifier circuit is the method that is most frequently used. Large antenna arrays and broadband antennas can be used to achieve this; however, doing so comes at the expense of size. An alternative would be to employ a harmonic rejection antenna or a low-pass filter between the rectifier and the antenna to prevent signals from being reradiated by the rectifier's nonlinear components.

1.5 RFEH ESTIMATION METRICS

The performance of a wireless energy harvesting (WEH) system is characterized by different parameters like duty cycle, operating distance, PCE, resonant Q factor, and so forth.

1.5.1 DUTY CYCLE

The proportion of time that the signal is ON for a given period of time is known as the duty cycle. It affects the energy available for radiofrequency energy harvesting. In Table 1.3, the duty cycle for different RF sources is provided.

1.6.3 ANTENNA

The broadband sleeve dipole antenna operating in the GSM band with an omnidirectional radiation pattern is used so that power can be received from all directions as well as over a wide frequency range. The size, shape, and weight of the antenna are optimized for better mobility. The antenna is vertically polarized with an omnidirectional radiation pattern. It operates from 800 MHz–3 GHz and has a gain of 1 dBi. The overall length of the antenna is 87 mm.

1.6.4 INDOOR MEASUREMENT

The Keysight N9917A FieldFox handheld microwave analyzer, along with the antenna, is used for signal strength measurement at various points indoors. RF mapping has been done inside the Centre for Research in Electromagnetics and Antennas (CREMA), Department of Electronics, CUSAT. The total area of the grid is 5 m x 2.5 m, which is marked on the floor, and each cell on the grid has an area of 0.5 m x 0.5 m. The measuring antenna is placed at each corner and center of the cell of the grid, and ambient received RF power is measured. The measurements are then averaged, and the result is shown in Figure 1.5. This plot reveals that the GSM communication bands show the highest power levels up to –53 dBm, and the lowest signal strength is about –66 dBm. It also reveals the presence of LTE, GSM 900, GSM 1800, and 2.5 GHz frequencies. Hence, GSM bands show the highest signal strength in the indoor environment. The 1760 MHz signal has the most strength at this point, which marked the power level of –53 dBm. This data suggests that for indoor communication or networking purposes, leveraging GSM frequencies would be most effective. Understanding these signal strengths aids in optimizing network coverage and connectivity within indoor spaces, ensuring reliable communication for various applications ranging from IoT devices to mobile phones. Moreover, identifying the strongest frequencies enables network operators to strategically deploy resources and infrastructure to enhance signal quality and overall user experience.

1.6.5 OUTDOOR ENVIRONMENT

To study the outdoor spectrum opportunities for RFEH, extensive drive tests have been conducted at 25 different locations inside the campus

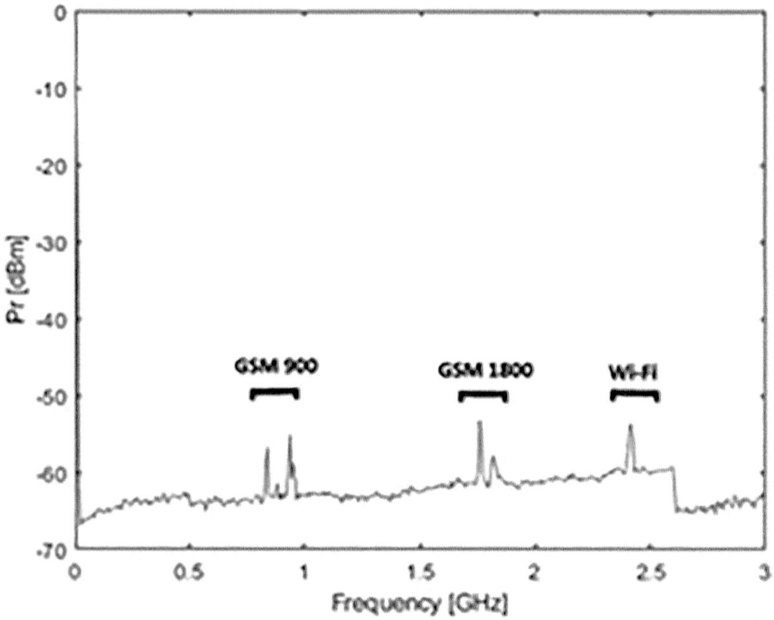

FIGURE 1.5 Signal strength obtained inside CREMA.

of CUSAT. Measurements have been taken using a measuring antenna mounted on the top of a car, which is connected to the Field Fox spectrum analyzer. Received RF power at all these 25 locations is measured several times, both during day and night conditions, and the received power values are averaged.

From Figures 1.6 a and b, it is clear that in the outdoor scenario, the available ambient RF signals show an impressive power level. All of the frequency bands under our discussion, namely GSM900, GSM1800, and also communication bands at 2100MHz and 2300MHz, along with the Wi-Fi signal band, show a good amount of power level outdoors and provide enough hope for the future of RFEH. Additionally, a strong radio frequency signal was observed in the 205 MHz–220 MHz area, which is found to be from nearby TV broadcasting stations. In the aforementioned outdoor drive tests, the maximum power received is almost –47dBm at the GSM900 band, exactly at 934MHz, both in day and night conditions. Though the maximum value of averaged power is -47dBm as per the driving test inside the campus, the scenario just outside the campus is a little more hopeful.

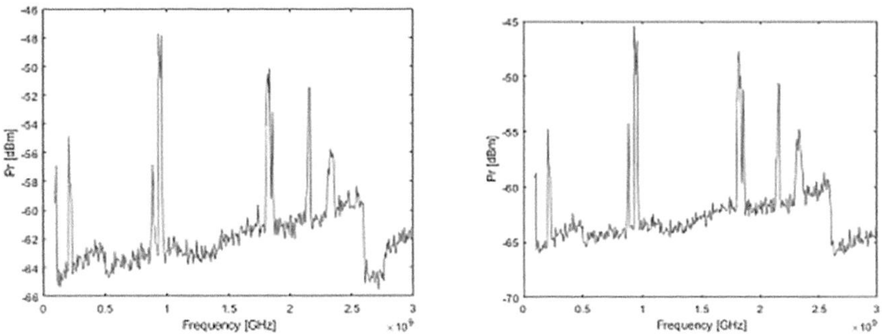

FIGURE 1.6 Signal strength obtained at outdoor campus during (a). Day (b). Night.

In comparison to the power emitted by the base station transmitters, the power received by mobile phones is quite small. In light of this, the base station transmitter is a better option than mobile transmitters. Applications where an RF harvester is powered by a mobile phone suggest that the phone and the harvester must be placed closer than five meters apart.

The three entities that provide the most to the CUSAT campus are 3G, GSM1800, and GSM900. The reflecting properties and multipath of each location account for the 20-fold reduction in the average received power level compared to the maximum power level at these bands. Therefore, in an urban or semi-urban setting, the distance to the transmitter does not yield the information needed to determine the power density that is accessible.

The received RF power level at 25 different outdoor locations has been measured, and the maximum received power is found to be near –45 dBm again in the GSM900 band. This supports the notion that GSM900 is the most promising frequency band for future RFEH systems, at least in our region. A broadcasting band of 205 MHz–220 MHz with a power level reaching up to –55 dBm was also noted. From the analysis of measurements done inside and just outside the campus, it can be concluded that the thick vegetation and tall, dense buildings on the campus may be significant reasons for the loss of RF power. Signal strengths observed from all these measurements, reaching up to –31 dBm, represent a feasible power level for RF Energy Harvesting. The indoor scenario consists of numerous Wi-Fi routers, and so obviously, the signal strength of Wi-Fi will be higher in those places.

1.7 ANTENNA

Advances in VLSI technology led to the creation of smaller devices, which necessitated the use of easily integrated, compact antennas. Reducing size comes at the expense of several elements, including efficiency, gain, bandwidth, and beamwidth. Wheeler (1947) provided the fundamental limit on size reduction. He states that the largest dimension of an electrically tiny antenna can be smaller than $\lambda/2\pi$. It lies in a sphere of radius a and ka less than 1. According to Chu, the antenna needs the following Q (Equation 1.8) in order to have the least amount of loss.

$$Q \geq \frac{1}{k^3a^3} + \frac{1}{ka} \tag{1.8}$$

The trade-off between antenna efficiency and size underscores the complexity of antenna design, necessitating the development of intelligent solutions. The formula Q = fr/BW, where fr represents the resonant frequency and BW denotes the bandwidth, elucidates this relationship. A higher quality factor (Q) indicates a narrower bandwidth but heightened efficiency, prioritizing resonance. Conversely, antennas with lower Q factors may offer broader bandwidths at the expense of efficiency. Understanding this interplay is crucial for optimizing antenna performance. By strategically balancing size, efficiency, and frequency coverage, engineers can create antennas that meet the diverse demands of modern communication systems.

The antenna Q factor decreases and the BW increases with decreasing size. The increase in BW will result in a decrease in gain because the gain-bandwidth product is constant. Therefore, there is a trade-off between each of these factors. In order to reduce the antenna's size, slits, and slots are employed. In this instance, fringing occurs at the ends of the slots, increasing the surface current path and resulting in a tiny form factor. However, the efficiency is degraded due to the impedance mismatch and narrow-band operation.

Slits reduce radiation resistance by compensating for ESA's reactive impedance through the use of slots. The mismatch in impedance between the load and the source therefore lowers efficiency. External circuits increase the form factor but decrease mismatch. A narrow-band frequency can achieve high efficiency, but the output voltage is limited to this range. A wideband design increases output by acquiring additional power. Broadband antenna design, in contrast to narrowband design, necessitates intricate circuitry to match the antenna with load impedance for changing input power and frequency.

The multiband input power is given by the following Equation 1.9.

$$P_M = \sum_{i=1}^{n} P_{fi} \tag{1.9}$$

The power at each tone is denoted by P_{fi}, the total multiband power P_M, and the number of tones is indicated by n. Thus, the power conversion efficiency may be found using the Equation 1.10.

$$\eta_M = \frac{P_M - P_{LO}}{P_M} \tag{1.10}$$

To improve energy harvesting, the antenna's polarization is crucial. Both vertical and horizontal polarization can be received by the circularly polarized antennas. Consequently, when the polarization mismatch loss decreases, the output power increases. Narrow bandwidth is produced by the techniques used to obtain circular polarization (CP). A few techniques for achieving CP are spur lines, corner truncation, stub loading, slits, and cross-shaped slots.

Rectennas' non-linear components produce fundamental frequency harmonics that reradiate onto the input side of the device and cause electromagnetic interference, which lowers performance. At the expense of gaining bulk, the low-pass filter placed between the antenna and rectifier absorbs reirradiations. This limitation can be overcome, albeit at the expense of larger dimensions, by rectifying the circuit and placing a low-pass filter between the antenna. Filtennas are some of the designs for antennas that have harmonic rejection. At the targeted frequency, harmonic suppression maximizes received power. Slit, stub, DGS, and other structural alterations are incorporated into the design of filters. An appropriate change is chosen based on the antenna's current distribution. Since they can only receive one frequency, many single-band antennas have a narrow bandwidth and can only gather modest power. Antennas with dual, quad, and broad bands receive power from several frequencies, increasing output. Multiport antennas are a smaller alternative to uniform arrays because they have a larger volume and area.

1.7.1 ANTENNA DESIGN

The E-shaped microstrip antenna is shown in Figure 1.7. The antenna is coaxially fed at (x, y). The antenna operation at 900 MHz is optimized by length L_1, width W_1, and inverted L of dimensions L_2 x W_2, which is optimized to obtain operation at 2.4 GHz. To achieve compactness at 900 MHz, a short is included at length W_3 from one end of the antenna. The substrate used is glass epoxy with a dielectric constant of 4.4 and a thickness of 1.6 mm. The antenna is coaxially fed at (x_1, y_1) with an SMA connector of 50 Ω impedance. The inner conductor is directly connected to the radiating patch, and the outer conductor is connected to the ground. The dimensions of the substrate are 50 mm x 47 mm. The radiating square patch has a dimension of 39 mm. The square patch has a dimension W_2. The simulations are done in CST software. The dimensions W_1, L_3, L_1, and W_2 are varied and optimized at 9 mm, 5 mm, 20 mm, and 39 mm to operate at 2400 MHz and 900 MHz. The values of these dimensions are given in Table 1.4.

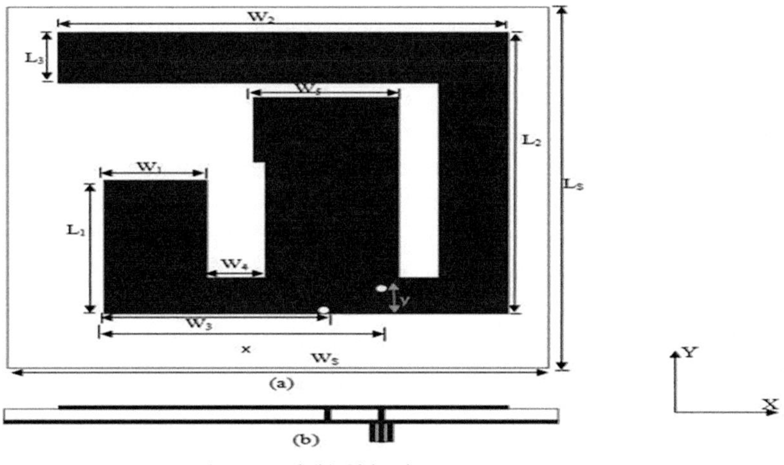

FIGURE 1.7 Antenna (a) Top and (b) Side view.

Source: Adapted from Ref. [6]

TABLE 1.4 Antenna Parameters

Parameter	L2	W3	W4	W5	W6	W7	X1	Y1
Dimension [mm]	29	19.5	5	3	11.5	12	23.5	4

Network Analyzer N9917A is used to fabricate and test the antenna. As seen in Figure 1.8, the measured findings and the calculated results correspond nearly perfectly, and the slight variations in the results are due to the losses in soldering as well as the slight differences in dielectric values. The results can be improved by machine soldering and fine-tuning.

By incorporating the short, the resonance is achieved at 900 MHz, and matching is improved at 2.4 GHz. The position of the shorting pin affects the antenna resonance, as the shorting posts are modeled as inductors. By varying its position, the equivalent impedance due to the inductance in parallel with the antenna equivalent circuit gets varied. The position in which maximum matching is obtained is selected by varying its position, and the optimum is obtained at x = 19.5 mm. The shorting position is varied along the x-coordinate from x = 19.5 mm in the positive x direction and also to the left of 19.5 mm. The antenna impedance and coax impedance matching decrease with a change of the short left of 19.5 mm, increasing return loss.

As seen in Figure 1.9(a), the antenna structure is positioned on each side of a cube to create a multiport antenna in order for it to accept radiofrequency radiation with orthogonal polarizations and from all directions. To create distinct polarizations, identical antenna elements are positioned on each of the cube's five sides in different orientations. However, because two antennas placed close to one another may interfere with each other's performance, antenna placement should be done carefully. The minimal interference is achieved here, as the isolation between each port is less than –30 dB. This level of isolation ensures that the antennas operate independently without significant mutual interference, maintaining their individual performance characteristics. Additionally, careful attention is paid to the physical layout and orientation of the antennas to further mitigate any potential interference effects. By optimizing the spacing and orientation of the antennas, engineers can maximize the effectiveness of each antenna element while minimizing cross-coupling and mutual coupling effects. This meticulous approach to antenna placement ensures reliable and efficient operation of the overall system, facilitating robust communication and signal reception across multiple polarizations.

The antennas are positioned at 45° on the top and bottom to produce circular polarization. The remaining antennas are arranged on all of the cube's sides in a vertical and horizontal alignment, which guarantees maximum power absorption. This is verified by the simulated 3D antenna pattern, which appears in Figure 1.9(b) when all of the antennas are powered simultaneously. The presence of ground at the back of the substrate reduces the device's back radiation at 2.35 GHz.

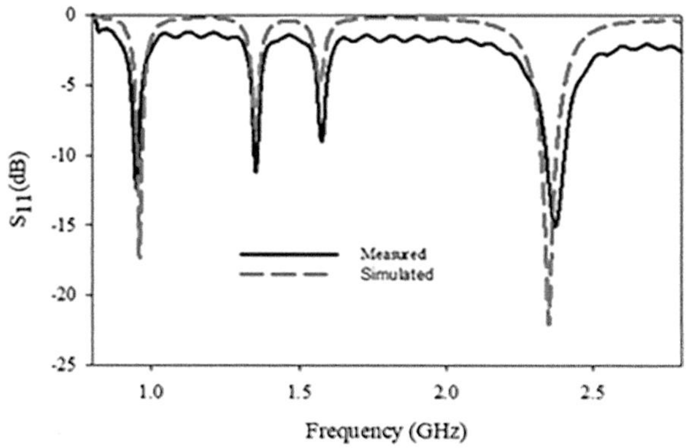

FIGURE 1.8 Reflection coefficient versus frequency.

Source: Adapted from Ref. [6]

Being able to accept power from all directions makes the multiport compact design suited for ambient RF energy gathering. The antenna is installed on both sides of the cube and runs at 2.35 GHz and 950 MHz; the antenna is 60 mm by 60 mm by 50 mm overall. The six-port antenna covers 360° and has good isolation of less than –30 dB. Because of its small size and low weight, it is perfect for powering low-energy Internet of Things devices. Because it is constructed on an FR4 substrate, it is an affordable choice. The device offers several benefits, including: (i) low cost; (ii) small size; (iii) 360° angular coverage; (iv) dual frequency operation; and (v) optimized design for optimum power gathering.

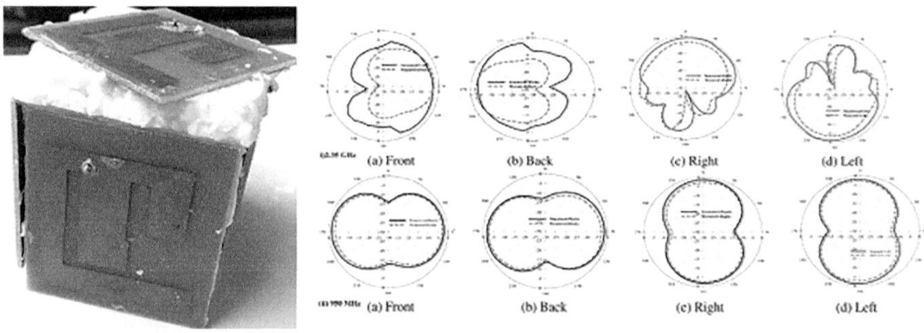

(a) (b)

FIGURE 1.9 (a) Cubical antenna (b) E plane radiation pattern.

Source: Adapted from Ref. [6]

1.8 RECTIFIER

The power leakage in low-power systems leads to insufficient energy. Hence, the maximum power transfer is achieved with the help of the addition of an impedance-matching network that matches the input impedance of the antenna with the impedance of the remaining circuit. Maximum power transfer is achieved when source impedance and load impedance are indistinguishable. When a mismatch between source and load impedance occurs, reflected power is produced in the circuit, which lowers the conversion efficiency. The impedance matching is obtained by adding reactive components between the source and the load.

The main configurations for matching networks are transformer, parallel coil, and LC network. The LC network and shunt inductor are used instead of the transformer in RFID tags and sensor networks due to economic reasons. The parallel coil is used to match high-impedance antennas like dipole antennas, whereas the LC network is suitable for small impedance antennas like Wi-Fi antennas. The transmission loss can be reduced, and the voltage gain can be increased by the impedance matching circuit. Hence, the antenna impedance is equal to the impedance seen by the antenna.

The basic voltage doubler is made up of a combination of diodes and capacitors, as shown in Figures 1.10 and 1.11. The output is produced only if the incoming signal voltage is greater than the barrier voltage. The doubler circuit increases the output voltage but not the power. However, due to the parasitic losses associated with the diodes and capacitors, the power decreases as the number of stages increases. So, a limitation is imposed on increasing the number of stages.

The Schottky diode is also known as a hot carrier diode. It got the name due to the higher energy electrons present in the n-type semiconductor compared to the electrons present in the metal. Here, a metal is made in contact with a lightly doped silicon material with no capacity to hold charges. Platinum, molybdenum, and chromium are some of the metals used for making Schottky diodes. The main advantage of the Schottky diode over the PN diode is that the effective voltage drop across the diode is very small, on the order of 200 mV, and also the fast-switching speed.

Different types of Schottky diodes are available with different threshold voltages. High cut-in voltage diodes are preferred for RFID applications where dedicated incident power is available, whereas low cut-in voltage diodes are preferred in the case of ambient energy harvesting, providing power for wearable electronics and IoT devices. The SMS 7630 diode is preferred for low incident power levels due to its low cut-in voltage.

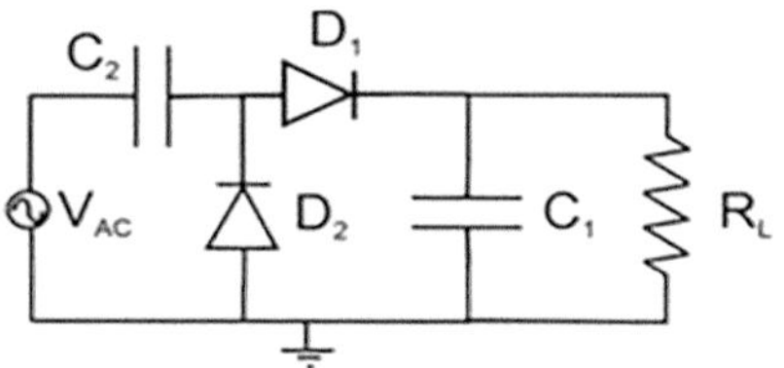

FIGURE 1.10 Single stage Villard multiplier.

Source: Adapted from Ref. [6]

1.8.1 LOW-COST RECTIFIER

The T-section matching circuit is used for achieving impedance match at 0 dBm input power. The T-section network is considered as two L-sections (Type 2 and Type 1) cascaded together. The first L-section transforms the resistance to real-valued impedance R_a, which is greater than source impedance R_s, and the second L-section transforms it back down to the original source impedance R_s. Thus, the net quality factor is greater than the single matching network, as shown in Equation 1.11.

$$R_m = \left(Q^2 + 1\right) R_{sma} \tag{1.11}$$

where Q factor is determined by bandwidth and operating frequency. R_{sma} is the smallest impedance of load impedance and R_s. After obtaining intermediate resistance R_m the impedance matching is obtained by the L matching networks as shown in Equation 1.12

$$Q_{se} = \sqrt{\frac{R_m}{R_L - 1}} = \frac{R_m}{X_{P1}} \tag{1.12}$$

where X_{p1} the inductive reactance of L_{p1} which is given by Equation 1.13.

$$L_{p1} = \frac{R_m}{\omega_0 Q_{se}} \tag{1.13}$$

The source impedance is 50 Ω and R_L = 3 Ω. The radial stub of width 7 mm is used to attain a nearly constant efficiency in the range 2 GHz–2.5

GHz. By the design equations the parameters of the rectifier circuit are calculated as in Table 1.5 and is shown in Figure 1.11.

TABLE 1.5 Parameter Values of Matching Circuit

Parameter	L1	C1	Length of Stub	C2	Radial Angle
Value	6 nH	0.1 pF	6 mm	100 pF	140°

These calculations provide crucial insights into the performance and efficiency of the rectifier, ensuring optimal energy harvesting from the RF signals. Table 1.5 outlines key parameters such as component values, operating frequencies, and expected performance metrics, offering a comprehensive overview of the rectifier's design specifications. Meanwhile, Figure 1.11 visually depicts the circuit layout, aiding in understanding the physical implementation and interconnection of components within the rectifier circuit. Together, these representations serve as valuable references for engineers and researchers involved in RF energy harvesting projects, facilitating accurate design implementation and performance evaluation.

Operating in the frequency range of 1.8–2.6 GHz, the rectifier with a matching circuit achieves an efficiency of 60% at 0 dBm input power. A voltage of 0.7 V was generated, as depicted in Figures 1.11 and 1.12. Because of the impedance matching network, the rectifier circuit functions with efficiency above 50% from 1.8 GHz–2.4 GHz, achieving 70% at that frequency.

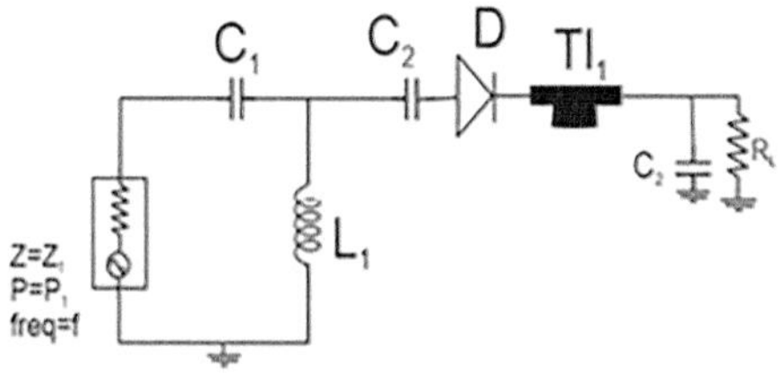

FIGURE 1.11 Rectifier circuit diagram.

Source: Adapted from Ref. [6]

Since the circuit is designed to operate at 0 dBm, fluctuations in input power beyond this point degrade rectifier performance because of the nonlinear element's impedance mismatch. However, it is appropriate for ambient energy harvesting because it offers an efficiency of 7% at –30

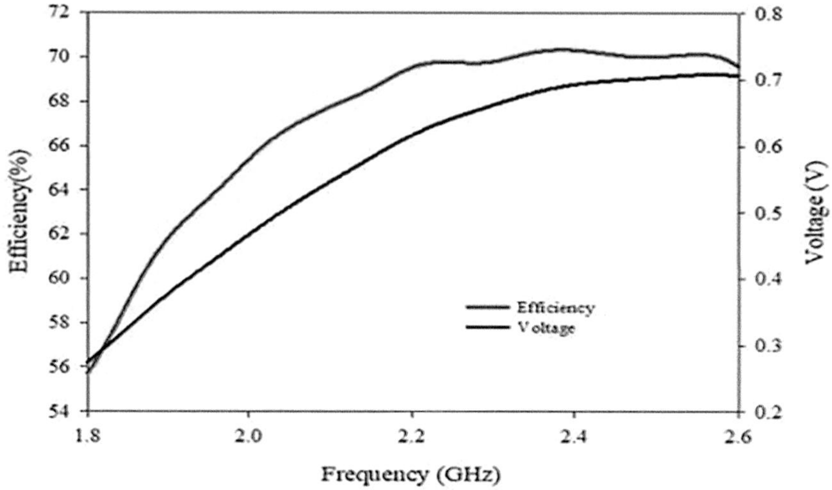

FIGURE 1.12 Efficiency versus frequency at 0 dBm input power.
Source: Adapted from Ref. [6].

dBm input power. The inexpensive rectifier, with an output voltage of 0.8 V and an efficiency of 62% while operating at 0 dBm input power, is intended to power Internet of Things devices. With an efficiency of 7% at –30 dBm input power, the rectifier is appropriate for harvesting ambient radio frequency energy.

1.9 CONCLUSION AND FUTURE WORK

Numerous tiny wireless systems capable of transmitting radiofrequency energy into the surrounding environment have been developed as a result of recent advancements in VLSI, embedded systems, and radio communications. Many researchers are interested in the rectenna as a means of recycling radiofrequency energy. In order to power sensor nodes wirelessly, RF energy harvesting systems have design-related problems that are identified in this work. The technologies used here help to solve the problems relating to rectenna design and the possibility of harvesting ambient RF power for powering the load.

To develop an effective rectenna system, the various components of RF energy harvesting systems need to be examined and optimized according to the application. There are three steps in the design of the RFEH system.

The system's power requirements need to be specified precisely. Power consumption in autonomous systems only rises when required, and we need to take into account the intervals between the active cycles and the average power used. Second, select the operating frequency based on the load. Dedicated RF energy harvesting is used for applications where the minimum power requirement is high, as well as where there is a direct line of sight between the transmitter and the receiver system. The sensor location, the form factor of the system, and the frequency of the signal used are decided in this stage. Finally, an efficient rectifier circuit must be selected and implemented to achieve increased power conversion efficiency and store the converted power in a suitable load. The effects of frequency and input power on power conversion efficiency were examined. The overall rectifier efficiency was increased by introducing novel techniques. Stub-matching circuits were used to optimize the rectifier for low input RF power operation. The rectifiers were inexpensive since they were built on FR4 substrates, and their efficiency was increased by using DC combining. Furthermore, a wide-band rectifier and a multiport small rectenna were developed and optimized to raise the PCE overall, enabling the harvesting of low incident power.

1.9.1 SCOPE OF FUTURE WORK

Millimeter waves, falling within the frequency range of 30 GHz–300 GHz with wavelengths spanning 1 mm–10 mm, constitute a critical segment of the electromagnetic spectrum. Positioned between microwaves and infrared radiation, these waves are instrumental in enabling high-speed communications. Recognizing their potential, the Federal Communications Commission (FCC) has advocated for the utilization of millimeter wave frequencies in 5G communication, primarily due to their ability to accommodate additional bandwidth, thereby facilitating faster video communication. One of the notable features of millimeter waves is their capability to support bandwidths of up to 2 GHz within bands reaching up to 100 GHz. Unlike lower-frequency bands, mmWave bands inherently possess ample bandwidth, obviating the need for additional band aggregation to achieve increased throughput. Moreover, owing to their narrow wavelengths, millimeter waves give rise to smaller cell sizes and higher cell density in communication.

These waves are also appealing for radiofrequency energy harvesting, in which a harvester extracts incident radiofrequency energy that can be used to power remote, low-power wireless sensor nodes. Due to diffraction and low penetrating power, buildings can cause obstructions. Attenuation by the atmosphere and precipitation is the main obstacle to mm-Wave communication. Figure 1.13 shows several millimeter wave energy harvesting sources. The development of diodes that operate beyond 30 GHz was the major challenge in designing mmWave rectennas for energy harvesting. Also, there are problems associated with printed circuit boards due to losses as well as etching tolerances. The main requirement of diode design is that its dimension must decrease as frequency increases to ensure that the impedance offered by junction capacitance is not too low during reverse bias. The junction capacitance is about 0.07 pF at 35 GHz, which corresponds to a reactance of 65 Ω. Other parameters to be considered are breakdown voltage, barrier voltage, and series resistance. The losses associated with these diodes are high, so reducing R_S appears to be a solution, but it in turn increases C_{j0}. So, a balance must be made between the two for the optimal performance of such diodes. The diode voltage is clamped at a lower voltage than its junction voltage because the energy storage in the inherent capacitors of these frequencies will determine its overall performance.

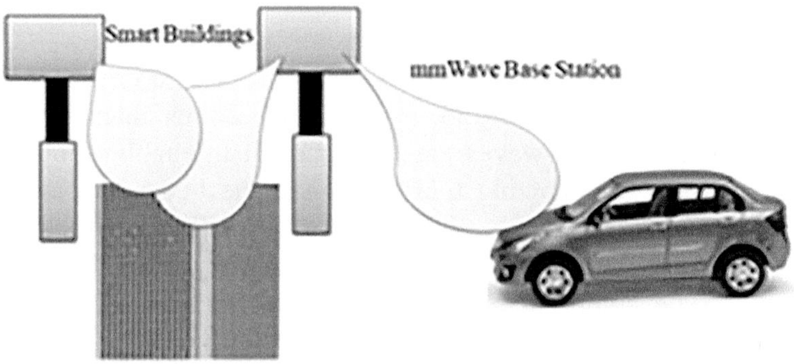

FIGURE 1.13 Millimeter wave energy harvesting.

Compared to MMIC technology, a hybrid retractable antenna made of an etched thin-film printed circuit board and a diode has a tolerance of 12 micrometers since it is more difficult to etch out the input filter and antenna at higher frequencies. At these frequencies, connecting diodes is challenging because it creates reactance at the joints that is uncorrectable at about 90 GHz. Thus, 35 GHz and higher applications are best suited for hybrid rectennas. Beyond 94 GHz, a fully integrated rectenna is useful. The devices' parasitic capacitances pose a major obstacle to the design of mmWave components. As frequency increases, the WPT efficiency is hindered by many practical challenges. The efficiency of the rectifiers at high frequency is hindered by multiple physical limits. Also, the conductive and dielectric losses associated with the antenna increase with increasing frequency, making low-loss substrates a great choice at these bands.

1.9.2 MMWAVE RECTENNA

Source-load impedance and matching topology determine the harvester PCE below the 6 GHz band. At mmWave frequencies, PCE is mostly determined by harmonic termination, parasitic effects, and diode packaging.

In microwave rectennas, distributed or lumped networks are used to achieve impedance matching. Here, minimizing S_{11} at the antenna interface is the primary means of maximizing PCE. Additional parts are required at mmWave frequencies to prevent high-frequency harmonics from propagating to the output, which could reduce system efficiency. The diode's non-linearities are more substantial and cause sufficient power to be created at higher harmonics, which can impact the DC output if they are not appropriately terminated. Furthermore, the primary problem at mmWave bands is the integration of all these components at this frequency. Therefore, several rounds are required to describe the diode parameters in a mmWave rectenna design. The CMOS IC technology is used to miniaturize rectennas.

1.9.3 MMWAVE ANTENNA

Due to high propagation loss at mmWave bands, high-gain antennas are used for maximizing harvested power. Table 1.6 compares antennas used for mmWave harvesting. Most of the designs are arrays with beamforming networks.

TABLE 1.6 Comparisons of Antennas Used Beyond 20 GHz

Rectenna	Antenna Structure	Substrate	Dimensions (mm)	Frequency (GHz)	Beamwidth
[24]	Folded dipole	65 nm CMOS	3.7 x 1.2 x 0.12	24	60
[25]	SIW	Duroid 5880	75 x 35	24	15
[26]	Coplanar Strip	R06002	10 x 13	21	50
[27]	4 x 4 array	RO3003	60 x 30	24	20
[28]	Lens array	LCP	55 x 87	28	80
[29]	Antipodal Vivaldi	Textile	16.5 x 9.5	24 – 40	50

Waveguides and waveguide-to-microstrip interfaces are used to develop mmWave rectennas having a gain greater than 15 dB. Circular polarization is preferred over linear polarization in WPT so that any angular misalignment between the transmitter and receiver doesn't affect the harvested power. Also, the waveguide components are costlier compared to planar antennas. The reflector-backed array antenna fabricated on a textile prescribed in Table 1.6 showed a higher gain compared to other microstrip patch array antennas. Additionally, on thinner substrates, microstrip antennas have a narrow bandwidth. However, microstrip antennas are easier to print in any shape using photolithography.

The substrate-integrated waveguide (SIW) development led to the advancement of planar antennas operating at mmWave bands. Their radiation losses are low due to the suppression of surface waves, and hence they show improved performance. SIW is demonstrated on conductive threads for developing wearable antennas with low loss and high isolation. Transmission lines up to 37 GHz are implemented using rigid threads, enabling textile-to-standard waveguide transition and facilitating easier integration of microstrip components like a rectifier.

Differentially balanced antennas, like dipoles, are preferred in CMOS-based energy harvesters as they can eliminate baluns required in single-ended ones. The position of the antenna operating at f1 and the storage capacitors are optimized so that mutual coupling with the antenna operating at the lower frequency f2 is minimized. An on-chip quarter-wave monopole operating at 71 GHz is directly connected to an inductively matched rectifier. The antenna can achieve a gain of 0 dB. In far-field energy receivers operating in mmWave bands, the Q factor of the matching circuit as well as antenna dimensions is equally significant.

1.10 CONCLUSION

The mmWave bands are a suitable platform for powering low-energy devices and the emergence of 5G. The various requirements and challenges for mmWave energy harvesting are reviewed in this chapter. The main requirements for efficient mmWave energy harvesting are summarized below.

- High gain and efficient antenna designs based on lossy substrates.
- Development of highly sensitive, impedance-matched rectifiers based on accurate packaging parasitic details.
- Antenna-rectifier integration with minimum loss.
- Low-cost antenna designs.

Rectifier and antenna design for the new mmWave rectennas are based on inexpensive substrates and high-efficiency packaging techniques. Printed antennas, flexible designs, and antennas in a package, together with the use of CMOS rectifiers, are innovative concepts for creating millimeter-wave radiofrequency energy harvesters.

KEYWORDS

- **antenna**
- **IoT devices**
- **power conversion efficiency**
- **rectenna**
- **rectifier**
- **RF energy harvesting**
- **sensors**
- **wireless energy harvesting**
- **wireless sensor nodes**

REFERENCES

1. Tesla, N. (1891). *U.S. Patent No. 454,622*. U.S. Patent and Trademark Office.
2. Yo, T. C., Lee, C. M., Hsu, C. M., & Luo, C. H. (2008). Compact circularly polarized rectenna with unbalanced circular slots. *IEEE Transactions on Antennas and Propagation, 56*(3), 882–886.

3. Notis, D. T., Liakou, P. C., & Chrissoulidis, D. P. (2004). Dual polarized microstrip patch antenna, reduced in size by use of peripheral slits. In *34th European Microwave Conference, 2004* (Vol. *1*, pp. 125–128). IEEE.
4. Divakaran, S. K., Krishna, D. D., & Nasimuddin. (2019). RF energy harvesting systems: An overview and design issues. *International Journal of RF and Microwave Computer-Aided Engineering, 29*(1), e21633.
5. Niotaki, K., Collado, A., Georgiadis, A., Kim, S., & Tentzeris, M. M. (2014). Solar/electromagnetic energy harvesting and wireless power transmission. *Proceedings of the IEEE, 102*(11), 1712–1722.
6. Divakaran, S., & Krishna, D. (2021). Dual-band multi-port rectenna for RF energy harvesting. *Progress in Electromagnetics Research C, 107*, 17–31.
7. Glaser, P. E. (1986). Power from the Sun: Its future. *Science, 162*, 956.
8. White, J. F. (2004). *LC Resonance and Matching Networks* (pp. 56–77). Wiley-IEEE Press.
9. Sze, S. M. (2008). *Semiconductor Devices: Physics and Technology*. John Wiley & Sons.
10. Kailuke, A. C., Agrawal, P., & Kshirsagar, R. (2013). Design and implementation of low power Dickson charge pump in 0.18 μm CMOS process. *A A, 15*(3).
11. Mandal, S., & Sarpeshkar, R. (2007). Low-power CMOS rectifier design for RFID applications. *IEEE Transactions on Circuits and Systems I: Regular Papers, 54*(6), 1177–1188.
12. Divakaran, S. K., Krishna, D. D., Nasimuddin, N., & Antony, J. K. (2019). Multipart compact antenna for RF energy harvesting. In *2019 IEEE Asia-Pacific Microwave Conference (APMC)* (pp. 1265–1267). IEEE.
13. Divakaran, S. K., Joy, L. M., & Krishna, D. D. (2018). Low-cost compact rectifier for RF energy harvesting. In *2018 IEEE Indian Conference on Antennas and Propagation (InCAP)* (pp. 1–4). IEEE.
14. Divakaran, S. K., & Krishna, D. D. (2017). An overview of compact antennas for RF energy harvesting. In *2017 IEEE International WIE Conference on Electrical and Computer Engineering (WIECON-ECE)* (pp. 47–50). IEEE.
15. Sun, H., Guo, Y. X., He, M., & Zhong, Z. (2013). A dual-band rectenna using broadband Yagi antenna array for ambient RF power harvesting. *IEEE Antennas and Wireless Propagation Letters, 12*, 918–921.
16. Balanis, C. (2005). *Antenna Theory: Analysis and Design*.
17. Xia, M., & Aissa, S. (2015). On the efficiency of far-field wireless power transfer. *IEEE Transactions on Signal Processing, 63*(11), 2835–2847.
18. Mikeka, C., & Arai, H. (2009). Design of a cellular energy-harvesting radio. In *2009 European Wireless Technology Conference* (pp. 73–76). IEEE.
19. Kuo, Y. W., Li, C. L., Jhang, J. H., & Lin, S. (2018). Design of a wireless sensor network-based IoT platform for wide-area and heterogeneous applications. *IEEE Sensors Journal, 18*(12), 5187–5197.
20. Hertz, H. (1893). *Electric Waves: Being Researches on the Propagation of Electric Action with Finite Velocity Through Space*. Dover Publications.
21. Islam, M. T., Ullah, M. H., Singh, M. J., & Faruque, M. R. I. (2013). A new meta surface superstrate structure for antenna performance enhancement. *Materials, 6*(8), 3226–3240.

22. Bülent, U. R. U. L. (2020). Triangle metamaterial-based energy harvester design for 5G frequency. In *2020 4th International Symposium on Multidisciplinary Studies and Innovative Technologies (ISMSIT)* (pp. 1–4). IEEE.
23. Anitha, V., Lakshmi, S. S. J., Lakshmi, M. L. S. N. S., Kumar, K. A., Teja, C. R., & Reddy, T. R. (2012). A circularly polarized stacked patch antenna using Polyflon substrate for wireless applications. *International Journal of Engineering Research & Applications, 2*, 903–905.
24. Wagih, M., Weddell, A. S., & Beeby, S. (2020). Millimeter-wave power harvesting: A review. *IEEE Open Journal of Antennas and Propagation, 1*, 560–578.
25. Nikandish, G., Staszewski, R. B., & Zhu, A. (2020). Breaking the bandwidth limit: A review of broadband Doherty power amplifier design for 5G. *IEEE Microwave Magazine, 21*(4), 57–75.
26. Koert, P., & Cha, J. T. (1992). Millimeter wave technology for space power beaming. *IEEE Transactions on Microwave Theory and Techniques, 40*(6), 1251–1258.
27. Moro, R., Agneessens, S., Rogier, H., Dierck, A., & Bozzi, M. (2015). Textile microwave components in substrate integrated waveguide technology. *IEEE Transactions on Microwave Theory and Techniques, 63*(2), 422–432.
28. Collado, A., & Georgiadis, A. (2013). 24 GHz substrate integrated waveguide (SIW) rectenna for energy harvesting and wireless power transmission. In *2013 IEEE MTT-S International Microwave Symposium Digest (MTT)* (pp. 1–3). IEEE.
29. Malik, B. T., Doychinov, V., Zaidi, S. A. R., Robertson, I. D., Somjit, N., Richardson, R., & Chudpooti, N. (2019). Flexible rectennas for wireless power transfer to wearable sensors at 24 GHz. In *2019 Research, Invention, and Innovation Congress (RI2C)* (pp. 1–5). IEEE.
30. Surender, D., Khan, T., Talukdar, F. A., De, A., Antar, Y. M., & Freundorfer, A. P. (2022). Key components of rectenna system: A comprehensive survey. *IETE Journal of Research, 68*(5), 3379–3405.
31. Aboualalaa, M., Mansour, I., & Pokharel, R. K. (2023). Energy harvesting rectenna using high-gain triple-band antenna for powering Internet-of-Things (IoT) devices in a smart office. *IEEE Transactions on Instrumentation and Measurement, 72*, 1–12.
32. Li, Z., Zhang, J. W., Wang, Y. C., He, D. P., & Zhang, C. (2023). Two-port five-band rectenna for ultra-low ambient RF energy harvesting. *IEEE Antennas and Wireless Propagation Letters, 22*(8), 1972–1976.

CHAPTER 2

Design and Analysis of Fractal Array Antenna for 5G mm Wave-FR2 Bands and Wireless Applications

ARUN RAJ* and DURBADAL MANDAL

Department of ECE, Computer Aided Electromagnetic Design Laboratory, National Institute of Technology (NIT), Durgapur, West Bengal, India

*Corresponding author

ABSTRACT

This chapter presents fractal array antennas designed for millimeter-wave applications within the 24-40 GHz spectrum. The design features modified fractal antennas arranged in an array, connected through a hexagonal ring feed line with dimensions of 1.6 mm height and 900 mm^2 substrate area. Simulation results show significant improvements in peak gain (5.9-6.9 dBi) for the fractal antenna designs with peak directivity of 6.89 and 7.13 dBi for dual-band radiation. Within the 5G NR bands, the antennas demonstrate enhanced radiation at 32 GHz and 36.6 GHz with specific angular orientations (Phi = 0°, Phi = 90°, Theta = 90°). The fractal slots concentrate current distribution, directing radiation at phi = 0° and Phi = 90°, achieving essential pencil beam properties for 5G millimeter-wave communication. The antennas achieve maximum return losses of 34 and 40 dB for dual-band radiation, with VSWR near unity at resonances. The array design covers 30.4-33.6 GHz bandwidths and 35.2-40 GHz, making them suitable for n259, n260, and n261 5G NR bands and ground-based radio navigation. The design's performance is validated through CST simulation and vector network analyzer measurements.

Wireless Technology: Advances and Innovations for Smart IoT Applications. Rajesh Kumar Dwivedi, Vinod Kumar Singh, and Raghav Dwivedi (Eds.)

DOI: 10.1201/9781998511495-2

2.1 INTRODUCTION

In order to meet the rising demands of wireless communication systems, antenna design has made significant strides recently [1–4]. Researchers have been investigating novel strategies motivated by natural and mathematical phenomena to pursue compact, practical, and multiband antennas [5–7]. Fractal geometry has become an intriguing source of inspiration for antenna design due to its intricate and self-similar patterns [8–11]. Incorporating elements influenced by the fractal carpet structure, this research study presents the design and analysis of a fed-slotted low-profile antenna for L Band short-range communication and wireless applications [12–16].

Fractal elements, well-known for their self-repeating patterns, have distinctive characteristics that make them an appropriate choice for antenna design [17–21]. The miniaturization, bandwidth, and radiation properties of antennas can all be improved through fractal geometry. The carpet fractal, which is distinguished by its interlocking square patterns, has the potential to produce effective radiation and a small antenna size [22–27].

All element antennas are joined with one corner edge to a 5 mm square corner edge in the proposed antenna design. This configuration incorporates slots to achieve the desired performance characteristics and allows for an optimized feeding length [28–31]. Improved performance across several frequency bands is anticipated by including the carpet fractal element in the antenna design [32–34].

In order to understand more about the current state of research in this area, a literature review was carried out with a focus on recent publications. In wireless applications, innovative fractal antennas have shown remarkable performance. Bharti and Sivia (2021) [1] introduced a circular ring element-based "multiband nested square-shaped ring fractal antenna," highlighting improved bandwidth and multiband operation. Kumar and Pharwaha (2020) [2] presented a "Hilbert curve fractal antenna" known for its compact form factor and multiband capability. Singhal and Singh (2020) [3] investigated an "elliptical monopole-driven super wideband fractal antenna," delivering excellent frequency responses across diverse frequencies. Addressing 5G needs, Karthikeya, A. Begaonkar, and Koul (2017) [4] devised a cost-effective "triple-band mmWave Sierpinski antenna," incorporating "uniplanar EBG" for high gain. These antennas collectively represent cutting-edge designs for wireless communication, spanning from multi-band rings to wideband monopoles, catering to various frequency and application requirements.

The present research concentrates on designing and analyzing a fed-slotted low-profile antenna for L Band short-range communication and wireless applications, building on the groundwork set by these studies. By achieving an improved bandwidth, a smaller size, and effective radiation characteristics, using the carpet fractal element aims to improve the antenna's performance [35–38]. The rest of the presented chapter is structured as follows. The methodology for designing antennas is fully described in Sections 2.3 and 2.4, along with the geometry and structure of the suggested antenna [39–41]. Sections 2.4 and 2.5 present the simulation setup and methodology for assessing the antenna's performance. The section also discusses the results and analysis of the simulation and highlights the benefits of the suggested antenna design.

The design aims to improve the antenna's bandwidth, size reduction, and radiation efficiency by utilizing the distinctive characteristics of fractal geometry. The remainder of this chapter delves into the specifics of the antenna design, simulation technique, outcomes, and analysis, illuminating the potential of the suggested design for real-world uses [42–49].

2.2 ANTENNA DESIGN

In the design of microstrip patch antennas, the width and length of a patch are determined by the following mathematical formulas. The following determines the "patch's width":

$$'\left((W)=()\left(\frac{(C)}{(2f\sqrt{\frac{(\varepsilon r+1)}{2}}}\right)\right)' \qquad (2.1)$$

where, the speed of light in "free space" is "c." $= 3\times10^{8}\,{}^{m}/_{s}$; "$f_r$"– "Resonate freq.," "$\varepsilon_r$" – "Relative permittivity"; The effective relative permittivity constant, "εreff," is calculated as follows: [14] accounts for the "effective relative permittivity," "ε_r":

$$'\left((W)=(\frac{1}{(2\mathrm{fr}\sqrt{\mu\varepsilon}))})(\frac{2}{((\in r)+1)}))\right)' \qquad (2.2)$$

$$`\left(\varepsilon reff) = (\frac{(\in r+1)}{2}) + (\frac{(\in r+1)}{2}) \times \left(1 + 12(\frac{(h)}{(w)}) \right)^{\frac{-1}{2}} \right)', \tag{2.3}$$

Where, "*h*" is "height of dielectric substrate" in "mm"; "w" is "Patch's width," 'ε_{reff}' – "Relative permittivity"; The ("Lreff") "effective" "length" is accounted as:

$$\left((\text{Lreff}) = (\frac{(c)}{(2f\sqrt{(\varepsilon r)})}) - (2.\Delta L) \right), \tag{2.4}$$

Equation (2.2) is used to compute "W," and the length's extension can be determined as follows:

$$'\left(\Delta L = (h) \times (\frac{((\varepsilon \text{reff}) + (0.3)) \times \left((\frac{W}{h}) + (0.264) \right)}{((\text{reff}) - (0.258)) \times \left((\frac{W}{h}) + (0.8) \right)}) \right)', \tag{2.5}$$

"Actual patch length" was determined as follows:

$$'\left(L = (\frac{1}{2fr\sqrt{(\varepsilon)}}) \times (\frac{1}{(\mu\varepsilon)}) - (2.\Delta L) \right), \tag{2.6}$$

'(L1 = (L + 6.h)' as well as "(W1 = (W + 6.h)" as "Effective length and width":

$$`\left(\text{Band width Percentage} = (\frac{2(f2 - f1)}{(f1 - f2)}) \times 100 \right)' \tag{2.7}$$

For "50 Ohms," the "feed point Position' can be accounted by,

$$\left(('\text{Rin } (y = y0)') = ('\text{Rin}(y = 0)') \cos\left(\frac{\pi}{L} y0 \right)^2 \right)' \tag{2.8}$$

Where "R_{in}(y = y0)" as "50 Ohms," Then "R_{in}(y = 0)" is roughly accounted as,

$$'('Zin') = ('1/Yin') = ('Rin')' = '('1/2G1')'$$

$$'\left((G1) = \begin{cases} (\frac{1}{90}\left(\frac{W}{\lambda}\right)^2)(W \ll \lambda) \\ \frac{1}{120}\left(\frac{W}{\lambda}\right))(W \gg \lambda) \end{cases} \right), \tag{2.9}$$

For Array factor "AF,"

$$"AF = \{Sin(N/2)\Phi\} / \{Sin(\Phi/2)\}' \tag{2.10}$$

where ("Φ" = (k d*Cos θ +β)), Where "*N*" the number of elements making the array, (k = "2Pi/wavelength").

2.3 DESIGN AND SIMULATION

2.3.1 SIMULATION AND ANALYSIS OF SIMPLE PATCH AND FRACTAL ANTENNAS

Figure 2.1 and Tables 2.1 and 2.2 display the iterative process to develop the recommended antenna. The basic design is a primary microstrip patch antenna with a patch height of 0.75 mm and a substrate height of 1.6 mm. The suspended substrate is 19.6 mm long and 19.6 mm wide. The next step is to create an array antenna with a 1.6 mm substrate height. Each element antenna in the array has a patch that is 10 mm long and 10 mm wide. The antenna's performance and attributes may be enhanced and improved using this iterative procedure. A range of antennas created using a fractal antenna element is shown in the figure mentioned in Figure 2.1. Three steps are involved in the design of the fractal antenna. First, the ground's measurements are 19.6 mm in length and 19.6 mm in width. A square slot with a 45° rotation and a 1 mm size is used in the design, which employs a Sierpinski carpet structure. Then, the same arrangement is repeated using 0.7 mm square slots. The feed point can be found at these coordinates: (5, 0.5). Moreover, the matching impedance is calculated using the standard design equations addressed in the antenna design section. The hexagonal feed with a 50-ohm matched feeding connector is applied, and the fractal antenna is designed, as shown in Figure 2.1.

TABLE 2.1 Design Parameters for Element Antenna

Antenna Description	Dimension of Element Antenna
Patch	10 × 10 mm^2
Height of substrate (h)	1.6 mm
Loss tangent (Tan δ)	0.002
Frequency range (f_r)	24–40 GHz
Patch used	line feed
Shape	Square Shape

TABLE 2.2 Different Notations of Design Parameters for Simple Patch Antenna

Notations	Ground Length × Width	Patch Length × Width	Feed Length and Width	Structure Shape
Dimension (mm)	19.6 × 19.6	10 × 10	4.8 and 1; 0.5×0.5; 1×1; 0.3 mm hexagonal ring	Sierpinski Carpet

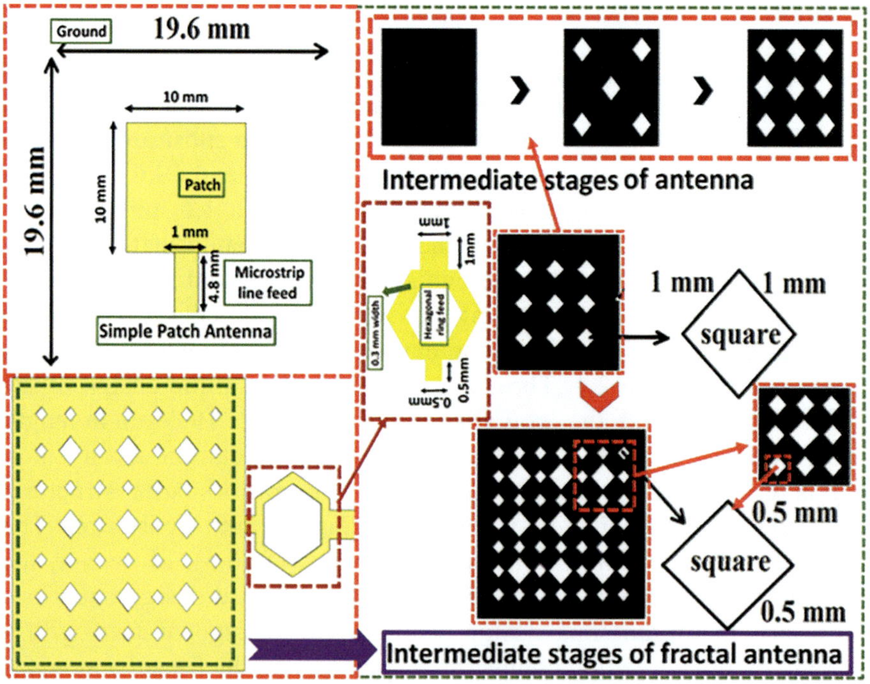

FIGURE 2.1 Fractal antenna through simple patch and with intermediate stages of formation.

Designing a fractal antenna entails a progressive refinement of its structure by incorporating simple patch elements with intermediate stages of formation, leveraging the principles of fractal geometry. This iterative process commences with a rudimentary patch antenna, serving as the foundation upon which intricate fractal patterns are gradually introduced along its edges. Examples of such patterns include the Koch snowflake and Sierpinski triangle, renowned for their self-replicating nature and complex geometric intricacies. When applied to antenna design, these fractal patterns confer several advantages, including enhanced bandwidth and reduced physical footprint. As the design iterations unfold, more sophisticated fractal geometries, such as the Menger sponge, are integrated, further augmenting the antenna's intricacy and performance potential. The synthesis of traditional antenna design principles with cutting-edge fractal geometry involves a meticulous balance of creativity and optimization. Utilizing electromagnetic simulation software, engineers meticulously fine-tune the antenna's geometry to ensure adherence to stringent performance specifications. This iterative refinement process is essential for achieving optimal antenna performance across a range of operational parameters, including frequency response, radiation pattern, and impedance matching. The resulting fractal antenna represents a harmonious convergence of traditional and contemporary design methodologies, epitomizing the synergistic relationship between form and function. Beyond its aesthetic appeal, the fractal antenna embodies a paradigm shift in antenna design philosophy, offering tangible improvements in performance and versatility. By harnessing the inherent properties of fractal geometry, engineers can unlock new avenues for innovation in antenna technology, paving the way for enhanced communication systems, remote sensing applications, and wireless networking solutions. In essence, the fractal antenna stands as a testament to the transformative power of interdisciplinary collaboration, driving advancements at the intersection of science, mathematics, and engineering.

The results and discussion section compares the Simple Patch Antenna (SPA) and the FA based on the findings depicted in Figures 2.1 and Tables 2.1–2.3. Specifically, Figures 2.2–2.5, and Table 2.3 highlight the results of SPA and FA in terms of currents and voltage concerning time, S_{11} (the return loss) versus frequency, considering different feeding modifications as a simple patch antenna. The graphs of the Smith chart, VSWR, admittance, impedance 3-dB beamwidth, gain, and front-back ratio are depicted in Figures 2.2, 2.3, and 2.4, respectively.

TABLE 2.3 Comparison of "Different Parameters" of Simple Patch and Fractal Antenna

Figure 2.1	Resonance Point (fr)	f1 (GHz)	f2 (GHz)	Bandwidth (GHz)	Return Loss (dB)
Simple patch	33	25.6	37.2	11.6	34
Fractal Antenna	28.556	25	30.6	5.6	34

The current-voltage signal depicted in Figure 2.2 provides a dynamic representation of the behavior of the fractal element antenna over time. This signal showcases a distinctive negative resistive characteristic, wherein the voltage and current initially remain at zero before undergoing rapid variation after a few nanoseconds and then returning to zero. This fluctuation pattern signifies the resonance behavior exhibited by the fractal element antenna, particularly in proximity to the 5G frequency band around 28 GHz. The swift variation of voltage and current, followed by a return to zero amplitude, underscores the antenna's remarkable resonance capabilities within the specified frequency range. This resonance behavior is crucial for ensuring efficient energy transfer and signal propagation in high-frequency applications such as 5G communication systems, where precise resonance tuning is essential for optimal performance and signal fidelity.

In Figure 2.3, a Smith chart is presented to analyze the impedance matching characteristics of both the simple patch antenna and the fractal antenna. The chart illustrates the matching impedance near unity, indicating

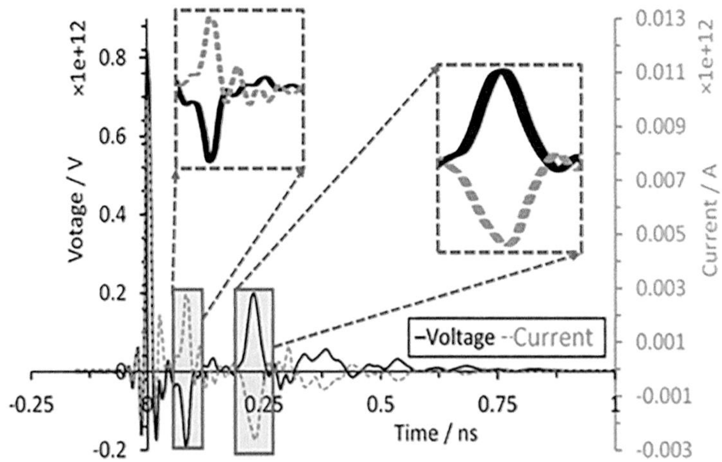

FIGURE 2.2 Current and Voltage graphs through the feeding port concerning fractal element antenna.

effective impedance matching between the antenna and the transmission line. The region around unity on the Smith chart suggests that the antenna's impedance closely matches the characteristic impedance of the transmission line, resulting in minimal signal reflection and efficient power transfer.

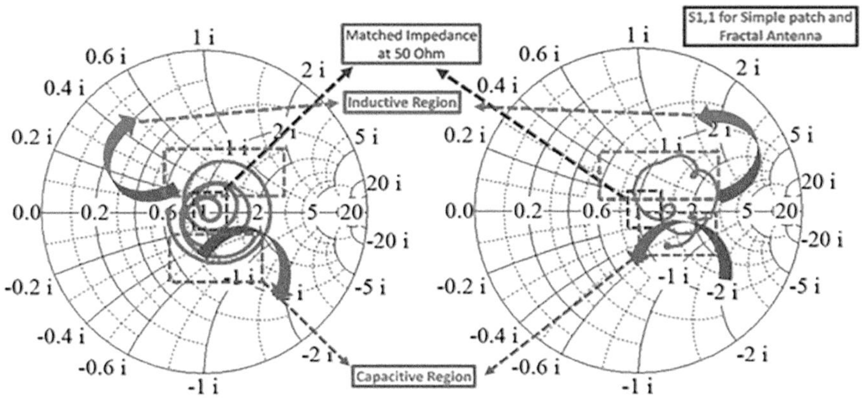

FIGURE 2.3 Smith chart for concerning simple patch and fractal antenna.

Figure 2.4 illustrates the reflection coefficient and VSWR (voltage standing wave ratio) concerning the "simple patch" and "fractal antennas." The simple patch antenna offers an 11.5 GHz broad bandwidth and a –34 dB reflection coefficient. However, enhancements involve integrating a carpet structure into the patch antenna and modifying the feed line. These modifications establish a 50 Ohm matching impedance. The fractal antenna achieves

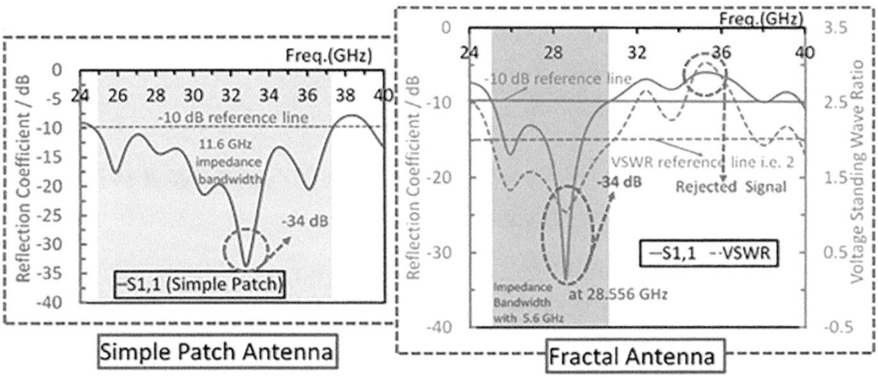

FIGURE 2.4 S_{11}, VSWR vs frequency graph for concerning simple patch and fractal antenna.

resonance with a distinct notch tailored for the 5G 28 GHz band. This setup ensures a wide 5.6 GHz bandwidth and an almost unity VSWR, effectively rejecting unwanted signal notches. In Figure 2.5, the fractal antenna showcases a peak gain of 5.2 dBi and a front-to-back ratio of 20.1 dB. Admittance and impedance graphs reveal a 50-ohm match at the resonant frequency of 28.556 GHz. Notably, the 3 dB beam width supports long-range communication. The isotropic sensitivity peaks at the fractal antenna's resonance points, indicating efficient performance for this configuration.

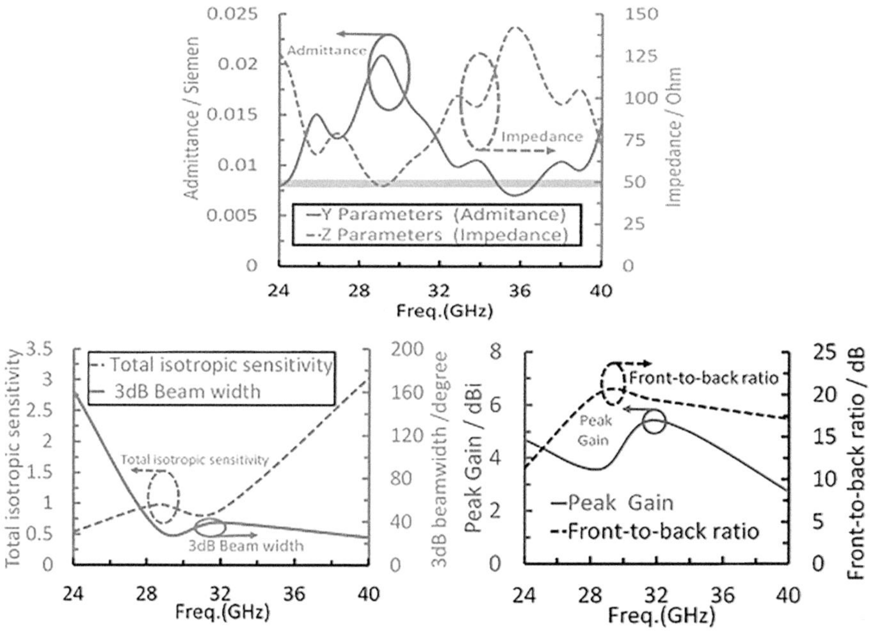

FIGURE 2.5 Y – Z parameters, 3 db beamwidth and sensitivity, peak gain and fbr concerning simple patch and fractal antenna.

2.3.2 SIMULATION AND ANALYSIS OF ARRAY ANTENNA WITH FRACTAL ELEMENT ANTENNA (FA)

Integrating a wind-shaped array antenna featuring a four-pointed star-defected ground structure (DGS) into a microstrip patch design offers many benefits. The utilization of the DGS significantly enhances the S_{11} parameter, leading to improved impedance matching and reduced signal reflection, consequently enhancing the overall antenna efficiency.

Matching impedance configurations are used in the array configuration, which involves connecting the edge corners of the element antennas to a central hexagonal feed. The first configuration uses a simple line feed connected to each element antenna, represented in Table 2.4 and Figure 2.6 (a). After that, the defected ground structure technique is implemented using the standard geometry of four-pointed star geometry. After this step, we repeat the same configuration four times. The unique wind-shaped array configuration optimizes port signal radiation by promoting better radiation pattern control and reduced side lobes, enhancing directivity and better signal coverage. Incorporating the DGS also contributes to heightened total isotropic sensitivity, increasing sensitivity across various angles and directions. Furthermore, this innovative design achieves notable peak gains, maximizing signal strength in the desired direction and bolstering overall communication performance. The results and discussion section compares the patch array antenna and the proposed array antenna using a defective ground technique to improve the resonance selectivity for the desired frequency range based on the findings depicted in Figure 2.7 and Table 2.4. Modifications on the patch and ground are depicted in Figures 2.6 (a, b, c, d) and 2.7 (a, b, c), respectively. Figures 2.8–2.10, and Table

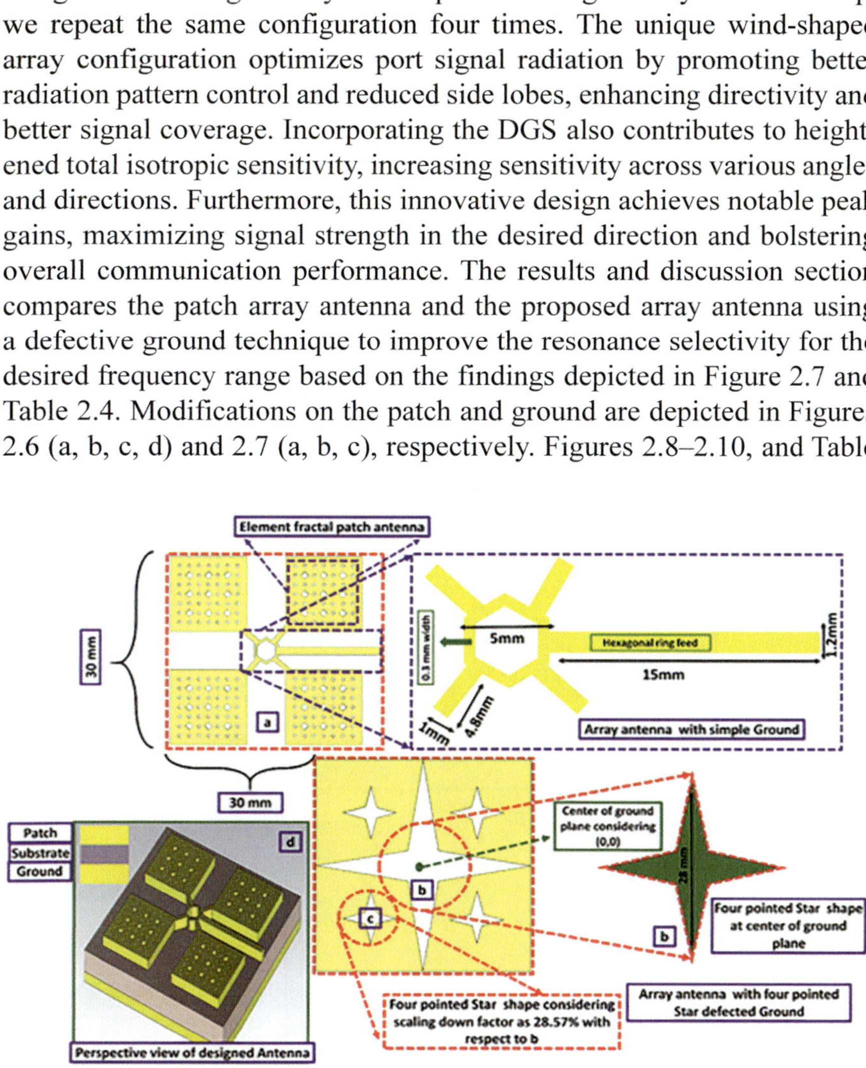

FIGURE 2.6 Array antenna 1 and 2 with defected ground depicted as a, and b, c d, view concerning front patch and ground structure.

2.5 highlight the results of SPA and FA in terms of currents and voltage concerning time, S_{11} (the return loss) versus frequency, considering different feeding modifications as a simple patch antenna. The graphs of the Smith chart, VSWR, admittance, impedance 3 dB beamwidth, gain, and front-back ratio are depicted in Figures 2.9 and 2.10, respectively.

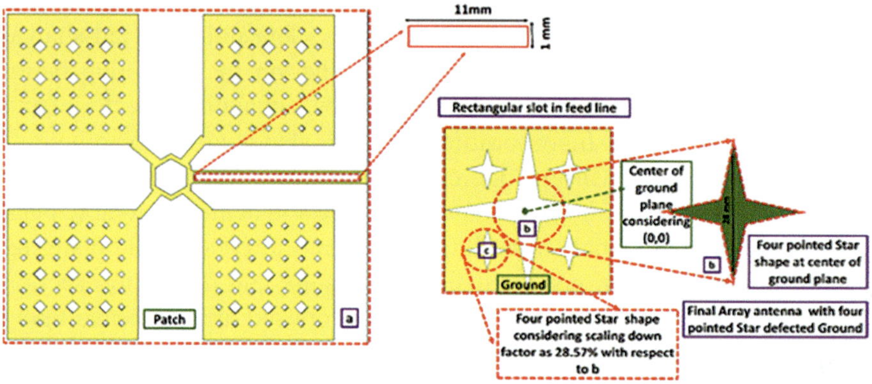

FIGURE 2.7 Final array antenna with defected ground depicted as a, b, c view concerning front patch and ground structure.

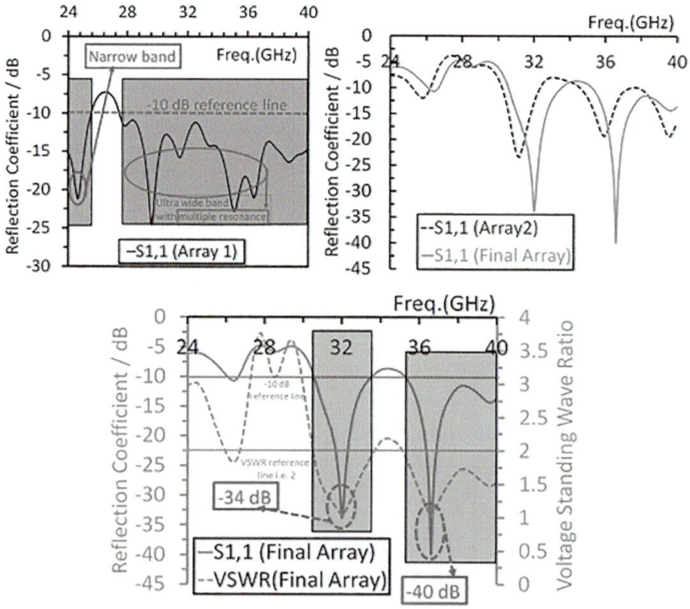

FIGURE 2.8 S_{11}, vs frequency and S_{11}, VSWR vs frequency graph for concerning array 1, 2 and final array antenna.

TABLE 2.4 Comparison of Different Parameters of Simple Patch and Fractal Antenna

Notations	Ground Length × Width'	Element Patch Length × Width	Feed Length and Width	Structure Shape	Ground Structure
Dimension (mm)	30 × 30	10 × 10	15 and 1.2; 0.3 mm hexagonal ring; 11 and 1 mm slot	Wind Mill shape	Four Pointed Star

The current-voltage signal concerning time, depicted in Figure 2.9, represents the negative resistance by which voltage and current vary from zero to variation after a few nanoseconds; it goes to zero, which shows the excellent resonance behavior for the fractal element antenna that resonates near 5G at 32 and 36.66 GHz, respectively. Figure 2.9 represents the Smith chart graph that depicts the matching impedance of about one, and the regions of inductive and capacitive behavior are depicted in Figure 2.9 mentioned for the simple patch antenna and the fractal antenna, respectively.

$$R(\ c) = \frac{Z(l) - Z(0)}{(Z(l) + Z(0)} = \frac{VSWR - 1}{VSWR + 2} \tag{2.11}$$

$$R(L) = -20 * \log_{10}(R(c)) \tag{2.12}$$

where $R(c)$ is reflection coefficient and $R(L)$ is return loss.

$$TRP = \int_0^{2\pi}\int_0^{\pi} R(\theta, \phi)\sin\theta d\theta d\phi \tag{2.13}$$

Equation 2.12 designates Watts [W] as the unit for Total Radiated Power (TRP). In the context of calculating TRP within an anechoic chamber, the process essentially involves computing the Effective Isotropic Radiated Power (EIRP) at various angles and then averaging it over the entire sphere. Therefore, by utilizing EIRP, we can ascertain the value of TRP.

$$TRP = \int_0^{2\pi}\int_0^{\pi} R(\theta, \phi)\sin\theta d\theta d\phi \tag{2.14}$$

Figure 2.10 and Table 2.5 illustrate the reflection coefficient and VSWR concerning the simple patch and fractal antennas. The simple patch antenna offers a wide range of bandwidth, featuring advanced 5G broad bandwidth. It achieves multiple resonance spots, delivering both narrowband and super wideband capabilities. The antenna's characteristics include a –22

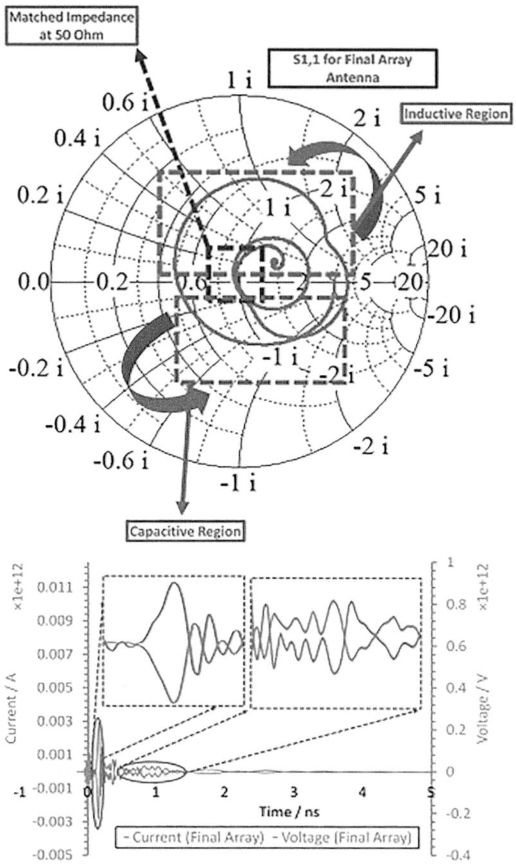

FIGURE 2.9 Smith chart, current/voltage graphs through the feeding port concerning proposed final array antenna.

dB and –24.85 dB reflection coefficient for broadband and narrowband, respectively. Additionally, using the first DGS method, the array antenna undergoes a resonance frequency shift to 31.6 GHz. This modification results in two distinct bands for the array antenna, featuring S_{11} values of –24.8 dB and –20 dB for their respective bands. In the proposed array, an optimized rectangular slot is integrated into the proposed antenna design. This innovation leads to selectively enhanced resonance, characterized by sharp resonance depth. Consequently, the reflection coefficients of the resonance attain –34 dB and –40 dB, respectively. Furthermore, with respect to the suggested array configuration, the frequency ranges addressed within the 5G NR spectrum are 30.4–33.6 GHz and 35.2–40 GHz. The proposed antennas are designed to operate across multiple bands, effectively encompassing various applications within the n259, n260, and n261 5G NR bands, including ground-based radio navigation applications. Additionally, the fabricated antennas cater to 5G applications with commendable matching performance.

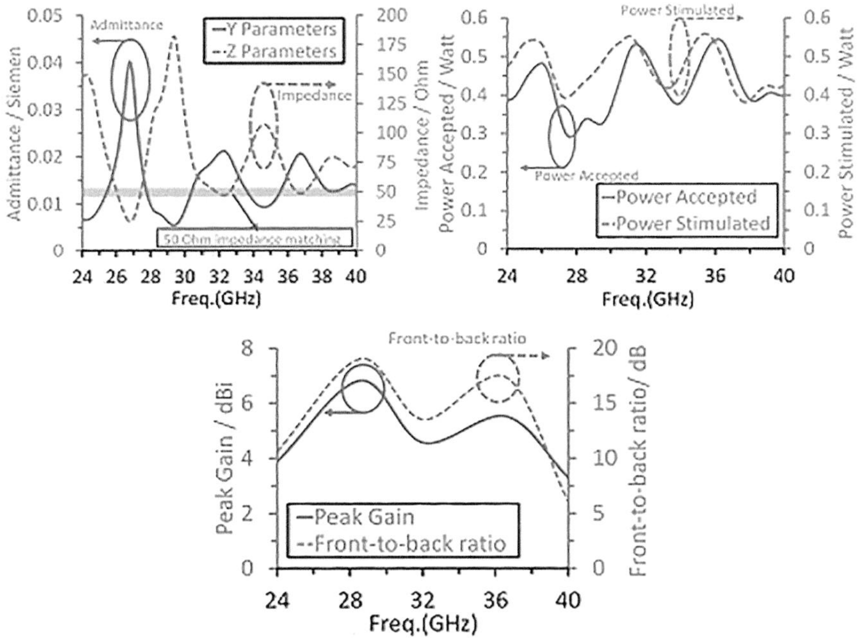

FIGURE 2.10 Y – Z parameters, Power accepted/stimulated, peak gain and front to back ratio vs frequency graph concerning proposed antenna.

TABLE 2.5 Comparative Study with Different Parameters of Proposed Array Antenna

Array	Size in mm^2	f_r (GHz)	f_1 (GHz)	f_2 (GHz)	Bandwidth (GHz)	Return Loss (dB)	VSWR
Array 1	30 × 30	24.7; 29.7; 35.2; 336.4	24; 27.6	25.8; 40	1.8; 2.4	21.2; 24.7; 22.75; 21	1.2; 1.12; 1.16; 1.19
Array 2 with DGS	30 × 30	31.2; 35.9; 39.59	30.2; 34.5; 37.5	32.34; 37.47; 40	2.14; 2.97; 2.5	23.5; 19.29; 19.3	1.14; 1.23; 1.21
Final Array	30 × 30	32; 36.6	30.4; 35.2	33.6; 40	3.2; 4.80	34; 40	1.04; 1.02

The antenna's compact size and ability to create sharp notches within the 5G NR band range make it a promising candidate for advanced 5G applications. One of the primary features of this fractal array antenna is its ability to determine the maximum main lobe direction as the beam direction, a crucial characteristic for optimizing signal transmission. The concept of a pencil beam is central to this research, given its relevance to achieving efficient long-range communication in 5G networks. Pencil beams offer advantages in covering extended distances, which is a vital requirement for catering to the demands of next-generation communication systems.

The significance of this research becomes evident through the electromagnetic patterns and resonances observed in the E–H plane. Figures 2.12 and 2.13 present these patterns, particularly highlighting resonant notches at 32 GHz and 36.66 GHz within the 24–40 GHz frequency range. These resonant notches are well-suited for facilitating communication within the 5G network bands. The E and H plane graphs provide insights into the direction of resonance, while the circled areas in red signify the intense movement of the electric and magnetic fields. These findings are made possible using CST software, underscoring the role of advanced simulation tools in antenna design and analysis.

The research presents comprehensive radiation pattern data through Figure 2.11 (a, b, c, d, and e). These figures offer a detailed perspective on the 2D power field radiation pattern and the 3D directivity planes. The specific angles of phi = 0°, phi = 90°, and theta = 90° are considered, and their respective resonances at 32 GHz and 36.6 GHz are highlighted. These visual representations further bolster the understanding of the antenna's radiation characteristics.

Tables 2.5–2.7 offer a quantitative analysis of the proposed antenna's properties, demonstrating its performance across various evaluation parameters. These parameters include resonating frequency, reflection coefficient, voltage standing wave ratio, bandwidth in GHz, fractional bandwidth percentage, front-to-back ratio, and minimum 3 dB beam width at resonance points. Such an assessment ensures that the antenna meets the desired specifications and can be effectively integrated into 5G communication systems.

Moreover, the research embodies a significant stride toward addressing the communication needs of the 5G era. The compact size, sharp notches for optimal bandwidth utilization, and emphasis on pencil beams collectively

TABLE 2.6 Comparative Study with Different Parameters of Proposed Array Antenna

Array	3 dB Beamwidth (Degree)	Bandwidth %	Linear Directivity (dBi)	Front to Back Ratio (dB)
Array 1	36, 20, 21.3, 19.4	7.2, 36.68	7.59, 8.99, 5.08, 8.72	14, 17.6, 7, 17.8
Array 2 with DGS	49.4, 15.4, 23.2	6.84, 8.25, 6.45	4.96, 6.61, 6.74	4, 8.5, 6
Final Array	11.2, 21.6	10, 12.76	6.89, 7.13	14, 20

TABLE 2.7 Experimental Range and Levels of Independent Variables

Properties	Proposed Antenna
Multi notch behavior	Yes
High Gain Directive	Yes
Wide band behavior	Yes
Compact size	Yes
Size Reduction	Yes
Pencil beam Behavior	Yes
MM wave 5G Applications	Yes

make the proposed fractal array antenna attractive for next-generation communication systems. The combination of theoretical analysis, simulation, and quantitative evaluation provides a comprehensive perspective on the antenna's potential and sets the stage for its practical implementation in advanced 5G applications (Figure 2.14).

2.4 FABRICATION PROCESS

The research focuses on the fabrication process of the fractal array antenna from a novel perspective. The antenna is constructed on a PCB board utilizing an FR4 substrate possessing an epsilon value of 4.3. Fabrication is efficiently executed using a standard PCB milling machine, the substrate maintaining a height of 1.6 mm. Figure 2.15 shows a distinct perspective of the structure on the PCB with the FR4 substrate, highlighting its enhanced compactness relative to the conventional coin. The innovation in this proposed fractal array antenna lies in its remarkable potential for substantial miniaturization and superior performance metrics. This antenna demonstrates heightened directivity, front-to-back ratio (FBR), gain, and bandwidth. By adopting this fabrication approach, the research showcases

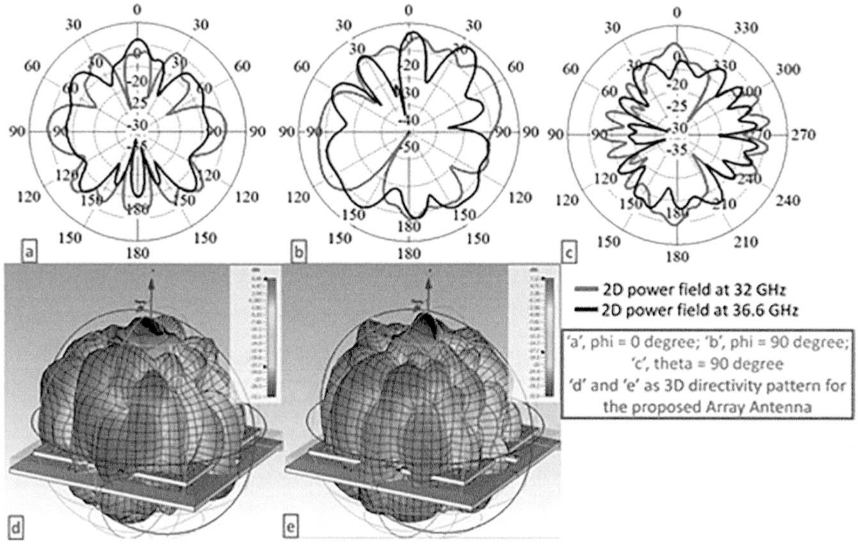

FIGURE 2.11 2D power field and 3D directivity pattern concerning proposed array antenna.

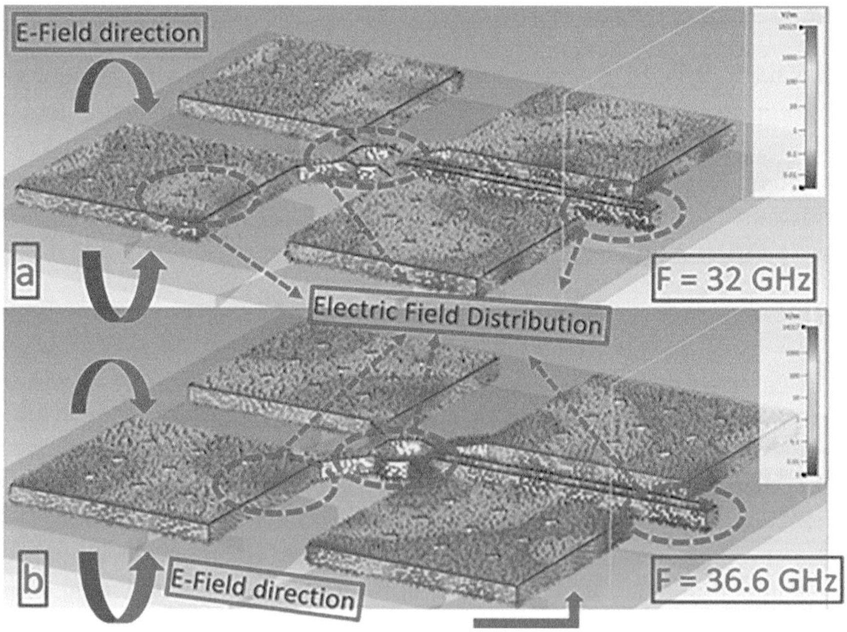

FIGURE 2.12 Electric field graph a and b concerning proposed array antenna.

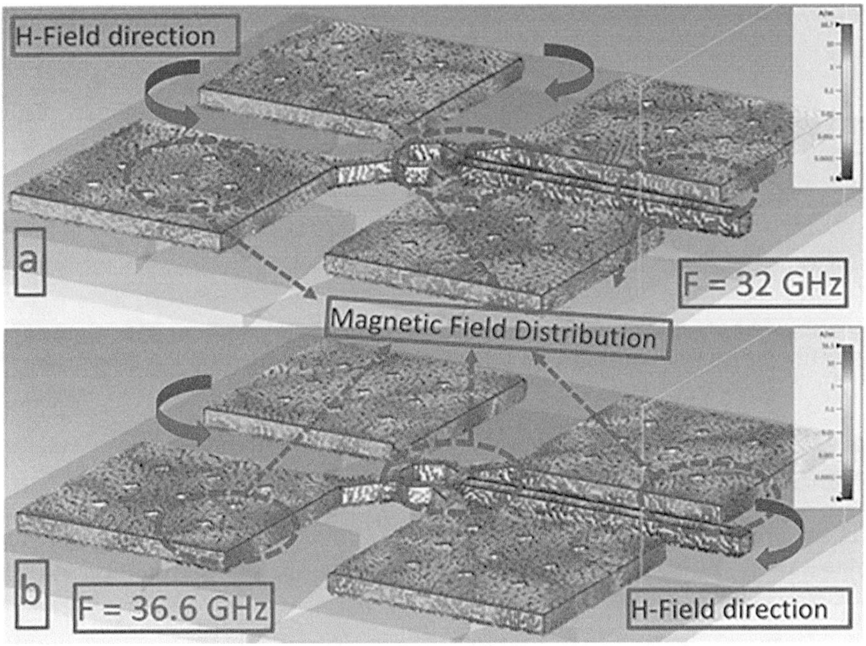

FIGURE 2.13 Magnetic field graph a and b concerning proposed array antenna.

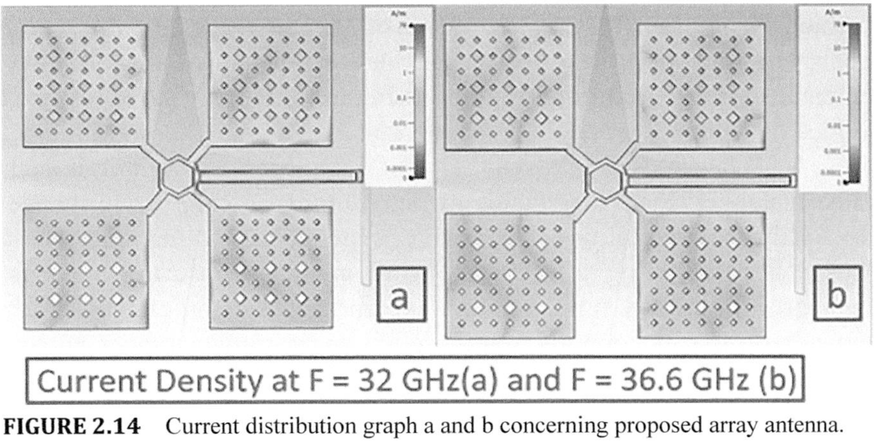

FIGURE 2.14 Current distribution graph a and b concerning proposed array antenna.

how the fractal array antenna is a competitive alternative in compact and high-performance antenna design.

FIGURE 2.15 Fabrication of proposed antenna front view, back view, and side view.

2.4.1 EXPERIMENTAL RESULTS WITH VECTOR NETWORK ANALYZER

The comprehensive investigation of the proposed antennas is presented through a rigorous research lens, emphasizing the meticulous design process and in-depth analysis carried out using the Vector Network Analyzer, as clearly illustrated in Figure 2.16. The obtained experimental results showcase remarkable resonance frequencies at 28.39, 32.04, and 36.66 GHz for the respective antennas. These frequencies are complemented by significant return loss values of 34, 18.84, and 23.5 dB, which decisively establish the effectiveness and efficiency of the proposed antenna designs. Additionally, the assessment of the VSWR revealed values of 1.06, 1.36, and 1.15 for the three antennas. The proximity of these VSWR values to the ideal value of 1 signifies outstanding impedance matching, a pivotal factor in determining antenna performance. To fortify the credibility of the results, a meticulous validation process was conducted using a vector network analyzer, ensuring a robust comparison between simulated and experimental outcomes. This study extends further by conducting a thorough comparative analysis against existing results, as succinctly compiled in Table 2.8. The outcomes of this comparison affirm the proposed antennas' promising performance across resonance frequencies, return loss, and VSWR, showcasing their resilience and competence when juxtaposed with other well-established outcomes. These findings underscore the significance of the designed antennas and their potential contributions to the field of antenna technology.

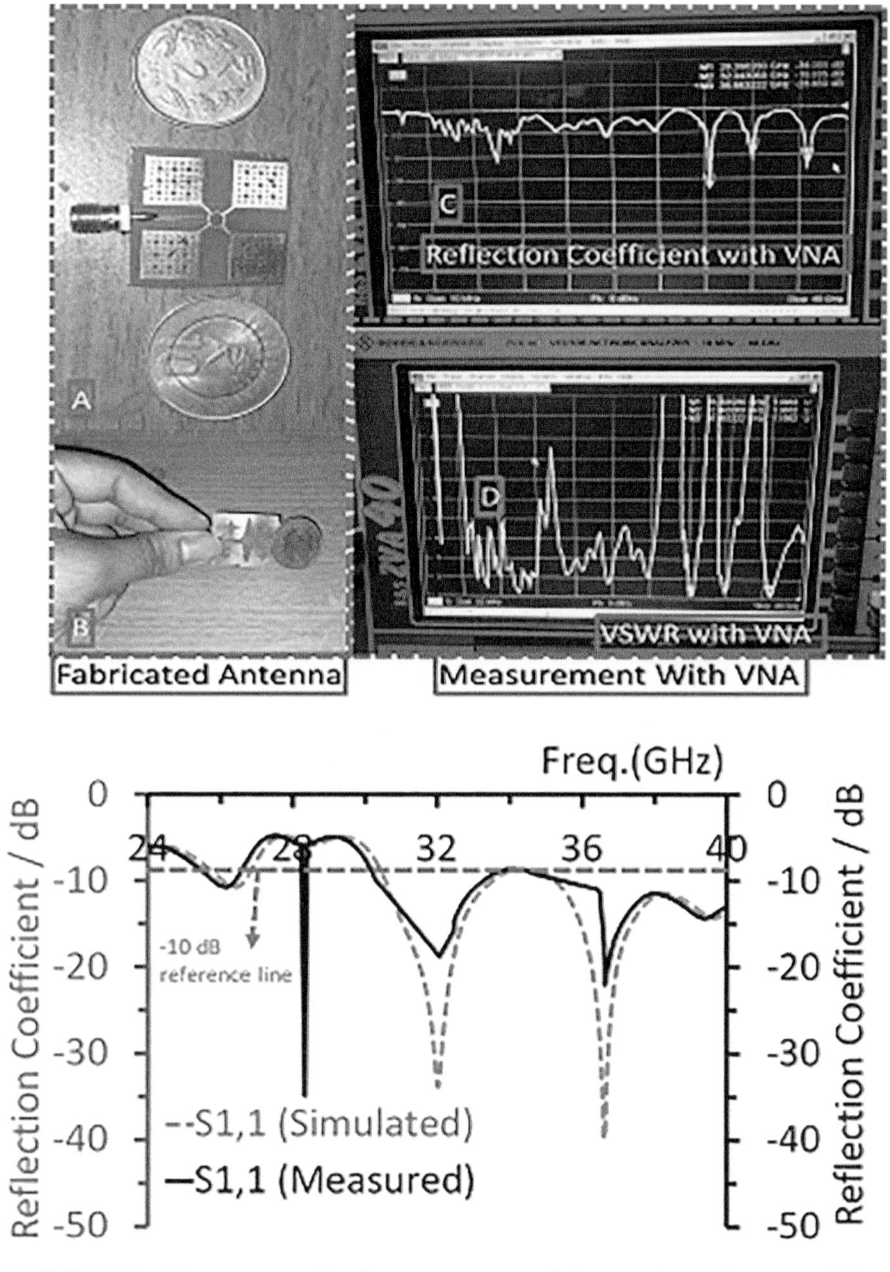

FIGURE 2.16 Measurement of various parameters and the experimental setup used for the fabricated prototype of the proposed antenna with a vector network analyzer (VNA).

TABLE 2.8 Comparative Analysis of the Results of the Proposed Antenna in Relation to Various Parameters in Previously Published Studies

References	Size (mm)	Peak Gain (dBi)	Bandwidth
[1]	36 × 32	NR	3.23 GHz
[22]	30 × 38	NR	680 MHz
[23]	38 × 38	NR	5 MHz
[24]	45 × 45	NR	200 MHz
[25]	40 × 46	NR	100 MHz
[28]	40 × 100	3.35	430 MHz
[29]	40 × 231	1.19	740 MHz
[30]	40 × 115	3.7	2.29 GHz
[31]	55 × 30	3	1.47 GHz
[32]	86.5 × 86.5	4.3	160 MHz
[33]	40 × 40	3.6	1.8 GHz
[34]	30 × 33	5.6	2.19 GHz
[35]	25× 20.5	4	3.31 GHz
[36]	32 × 15	2.73	1.93 GHz
[37]	40 × 40	5.5	4.7 GHz
[38]	40 × 26	5	3.20 GHz
[39]	50 × 30	5	1.67 GHz
[40]	52 × 52	2.3	1.04 GHz
[41]	57.9×57.9 53.6 × 52	6.2, 7.1, 7.3, 7.9	200 MHz, 330 MHz, 54 MHz, 80 MHz, 250 MHz
[42]	32 × 12	NR	1.41 GHz
Proposed Array Antenna	30 × 30	6.9	3.2 GHz, 4.80 GHz (Applications covered 5G mm wave range)

2.5 CONCLUSION

This research presents a groundbreaking contribution to antenna design by introducing fractal array antennas with versatile multiband capabilities. These antennas operate effectively as array elements, covering the mm-wave spectrum of 24–40 GHz. The antennas exhibit substantial performance improvements by integrating modified fractal patterns and a hexagonal ring feed line. The simulations authenticate a substantial 6.9 dBi peak gain increase, confirming the efficacy of the proposed design. Notably, exceptional radiation patterns are observed within 5G NR bands, particularly at 32 GHz and 36.6 GHz, with 34 dB and 40 dB return loss; however, prototype testing demonstrates resonance close to theoretical predictions at 32.04 and 36.66 GHz with 18.8 and 23.5 dB return loss, respectively. The concentration of current distribution on fractal slots highlights their superior radiating capabilities, crucial for advanced 5G mm-wave communications' pencil beam attributes. Furthermore, the array's extensive bandwidth coverage across 5G NR ranges, encompassing n259, n260, and n261 bands, underscores its applicability for diverse ground-based radio navigation and 5G technologies. Both simulation and real-world measurements affirm the pragmatic potential of these fractal array antennas, contributing significantly to the evolution of 5G capabilities.

ACKNOWLEDGMENT

We would like to thank SERB, Department of Science and Technology, Government of India (Sanction Order No: EEQ/2021/000700; Dated 04 March, 2022) for providing the fund that has been required to pursue this research.

KEYWORDS

- **fractal antennas**
- **fractal array antenna**
- **microstrip patch antenna**
- **resonance frequencies**
- **simulation**
- **wireless applications**

REFERENCES

1. Bharti, B., Bharti, G., & Sivia, J. S. (2021). A design of multiband nested square-shaped ring fractal antenna with circular ring elements for wireless applications. *Progress In Electromagnetics Research C, 108*, 115–125.
2. Kumar, A., & Pharwaha, A. P. S. (2020). Development of a modified Hilbert curve fractal antenna for multiband applications. *IETE Journal of Research, 68*(3), 1–10.

3. Singhal, S., & Singh, A. K. (2020). Elliptical monopole-based super wideband fractal antenna. *Microwave and Optical Technology Letters, 62*(3), 1324–1328.
4. Karthikeya, G. S., Abegaonkar, M. P., & Koul, S. K. (2017). Low-cost high-gain triple-band mmWave Sierpinski antenna loaded with uniplanar EBG for 5G applications. In *2017 IEEE International Conference on Antenna Innovations & Modern Technologies for Ground, Aircraft and Satellite Applications (iAIM)* (pp. 1–5). IEEE.
5. Singh, A., & Singh, S. (2016). Design and optimization of a modified Sierpinski fractal antenna for broadband applications. *Applied Soft Computing, 38*, 843–850.
6. Ali, T., Subhash, B. K., & Biradar, R. C. (2018). A miniaturized decagonal Sierpinski UWB fractal antenna. *Progress In Electromagnetics Research C, 84*, 161–174.
7. Choukiker, Y. K., & Behera, S. K. (2017). Wideband frequency reconfigurable Koch snowflake fractal antenna. *IET Microwaves, Antennas & Propagation, 11*(2), 203–208.
8. Dwivedi, R., Srivastava, D. K., & Singh, V. K. (2023). A novel miniaturized triangular slotted and DGS-based inverted circular key-shaped textile antenna with enhanced bandwidth for C, WLAN, RAS, N795G band application. *Iranian Journal of Science and Technology, Transactions of Electrical Engineering, 48,* 17–36. https://doi.org/10.1007/s40998-023-00651-4.
9. Singh, V. K., Bangari, N., Ali, Z., Vyas, A., Verma, R. K., & Saxena, A. (2023). Parachute shape ultra-wideband wearable antenna for remote health care monitoring. *International Journal of Communication Systems, 36*(10), e5488. https://doi.org/10.1002/dac.5488.
10. Yadav, A., Singh, P., Verma, R. K., & Singh, V. K. (2023). Design and comparative analysis of circuit theory model-based slot-loaded printed rectangular monopole antenna for UWB applications with notch band. *International Journal of Communication Systems, 36*(3), e5390. https://doi.org/10.1002/dac.5390.
11. Dwivedi, R., Meetkamal, & Dwivedi, R. K. (2021). Engineered nanoparticles for prevention against COVID-19 infection. *Advanced Materials Research, 1166*, 41–55. https://doi.org/10.4028/www.scientific.net/amr.1166.41.
12. Hashmi, S. N. A., Singh, V. K., Dwivedi, R. K., Pathak, R. S., & Tiwari, R. (2022). Tri pentagon slotted antenna using jeans material WLAN/HiperLAN application. In *2022 International Conference on Advancements in Smart, Secure and Intelligent Computing (ASSIC)* (pp. 1–4). IEEE. https://doi.org/10.1109/ASSIC55218.2022.10088371.
13. Sharma, D., Singh, V. K., Vyas, A., Kumar, N., & Dwivedi, R. K. (2023). Semi-circle slotted triangular shape antenna using flexible material. In *2022 International Conference on Advancements in Smart, Secure and Intelligent Computing (ASSIC)* (pp. 1–5). IEEE. https://doi.org/10.1109/ASSIC55218.2022.10088293.
14. Saha, R., Das, A., Mandal, D., & Kar, R. (2023). An optimal circular antenna array design for an efficient 5G communication system using krill herd optimization. In D. K. Sarkar, P. K. Sadhu, S. Bhunia, J. Samanta, & S. Paul (Eds.), *Proceedings of the 4th International Conference on Communication, Devices and Computing (ICCDC 2023)* (Vol. 1046). Springer. https://doi.org/10.1007/978-981-99-2710-4_51.
15. Kumar, S., Ram, G., Mandal, D., & Kar, R. (2023). Optimal design of time-modulated linear and circular antenna array for feed network efficiency using PSO. In *2023*

International Conference on Recent Advances in Electrical, Electronics, Ubiquitous Communication, and Computational Intelligence (RAEEUCCI) (pp. 1–5). IEEE. https://doi.org/10.1109/RAEEUCCI57140.2023.10134328.

16. Bhatia, S. S., & Sivia, J. S. (2019). On the design of fractal antenna array for multiband applications. *Journal of The Institution of Engineers (India): Series B, 100*(5), 471–476.
17. Sran, S. S., & Sivia, J. S. (2020). ANN and IFS-based wearable hybrid fractal antenna with DGS for S, C, and X band application. *AEU-International Journal of Electronics and Communications, 127*, 153425.
18. Rani, R. B. (2020). A compendious review on fractal antenna geometries in wireless communication. In *2020 International Conference on Inventive Computation Technologies (ICICT)* (pp. 888–893). IEEE.
19. Aravindraj, E., Nagarajan, G., & Kumaran, R. S. (2020). Design and analysis of recursive square fractal antenna for WLAN applications. In *2020 International Conference on Emerging Trends in Information Technology and Engineering (ic-ETITE)* (pp. 1–5). IEEE.
20. Sivia, J. S., Singh, A., & Kamal, T. S. (2013). Design of Sierpinski carpet fractal antenna using artificial neural networks. *International Journal of Computer Applications, 68*(8), 5–10.
21. Pandeeswari, R., & Raghavan, S. (2015). A CPW-fed triple-band OCSRR embedded monopole antenna with modified ground for WLAN and WiMAX applications. *Microwave and Optical Technology Letters, 57*(10), 2413–2418.
22. Ahmad, B. H., & Nornikman, H. (2013). Fractal microstrip antenna with Minkowski Island split ring resonator for broadband application. In *2013 IEEE International RF and Microwave Conference (RFM)* (pp. 214–218). IEEE.
23. Dangkham, P., & Phongcharoenpanich, C. (2015). A compact split ring resonator antenna on paper-based for UHF RFID passive tag. In *2015 IEEE Conference on Antenna Measurements & Applications (CAMA)* (pp. 1–4). IEEE.
24. Sharma, V., Lakwar, N., Kumar, N., & Garg, T. (2018). Multiband low-cost fractal antenna based on parasitic split ring resonators. *IET Microwaves, Antennas & Propagation, 12*(6), 913–919.
25. Pandeeswari, R., & Raghavan, S. (2015). Microstrip antenna with complementary split ring resonator loaded ground plane for gain enhancement. *Microwave and Optical Technology Letters, 57*(2), 292–296.
26. Rahimi, M., Maleki, M., Soltani, M., Arezomand, A. S., & Zarrabi, F. B. (2016). Wideband SRR-inspired slot antenna with circular polarization for wireless applications. *AEU-International Journal of Electronics and Communications, 70*(9), 1199–1204.
27. Rao, M. V., Madhav, B. T. P., Anilkumar, T., & Nadh, B. P. (2018). Metamaterial-inspired quad-band circularly polarized antenna for WLAN/ISM/Bluetooth/WiMAX and satellite communication applications. *AEU-International Journal of Electronics and Communications, 97*, 229–241.
28. Ji, J. K., Kim, G. H., & Seong, W. M. (2010). Bandwidth enhancement of metamaterial antennas based on composite right/left-handed transmission line. *IEEE Antennas and Wireless Propagation Letters, 9*, 36–39.

29. Zhang, Q., & Gao, Y. (2017). Compact low-profile UWB antenna with characteristic mode analysis for UHF TV white space devices. *IET Microwaves, Antennas & Propagation, 11*(11), 1629–1635.
30. Rayno, J. T., & Sharma, S. K. (2012). Wideband frequency-reconfigurable spirograph planar monopole antenna (SPMA) operating in the UHF band. *IEEE Antennas and Wireless Propagation Letters, 11*, 1537–1540.
31. Indhu, K. K., & Mohanan, P. (2011). Compact broadband top-loaded planar monopole antenna for low-frequency applications. In *2011 Indian Antenna Week (IAW)* (pp. 1–3). IEEE.
32. Andriamiharivolamena, T., Lemaître-Auger, P., Tedjini, S., & Tirard, F. (2015). Compact planar monopole antenna for wearable wireless applications. *Comptes Rendus Physique, 16*(9), 851–861.
33. Rad, M. A., Soheilifar, M. R., & Zarrabi, F. B. (2019). Compact microstrip antenna based on fractal metasurface with low radar cross-section and wide bandwidth. *AEU-International Journal of Electronics and Communications, 98*, 74–79.
34. Wu, C. M. (2007). Wideband dual-frequency CPW-fed triangular monopole antenna for DCS/WLAN application. *AEU-International Journal of Electronics and Communications, 61*(9), 563–567.
35. Shanmuganantham, T., Balamanikandan, K., & Raghavan, S. (2008). CPW-fed slot antenna for wideband applications. *International Journal of Antennas and Propagation, 2008*.
36. Cui, Y. Y., Sun, Y. Q., Yang, H. C., & Ruan, C. L. (2008). A new triple-band CPW-fed monopole antenna for WLAN and WiMAX applications. *Progress In Electromagnetics Research M, 2*, 141–151.
37. Ray, K. P., Thakur, S. S., & Deshmukh, R. A. (2012). Wideband L-shaped printed monopole antenna. *AEU-International Journal of Electronics and Communications, 66*(8), 693–696.
38. Chitra, R. J., & Nagarajan, V. (2013). Double L-slot microstrip patch antenna array for WiMAX and WLAN applications. *Computers & Electrical Engineering, 39*(3), 1026–1041.
39. Suma, M. N., Bijumon, P. V., Sebastian, M. T., & Mohanan, P. (2007). A compact hybrid CPW-fed planar monopole/dielectric resonator antenna. *Journal of the European Ceramic Society, 27*(8, 9), 3001–3004.
40. Chen, L., Ren, X., Yin, Y. Z., & Wang, Z. (2013). Broadband CPW-fed circularly polarized antenna with an irregular slot for 2.45 GHz RFID reader. *Progress In Electromagnetics Research Letters, 41*, 77–86.
41. Ullah, H., & Tahir, F. A. (2020). A novel snowflake fractal antenna for dual-beam applications in 28 GHz band. *IEEE Access, 8*, 19873–19879.
42. Nhlengethwa, N. L., & Kumar, P. (2021). Fractal microstrip patch antennas for dual-band and triple-band wireless applications. *International Journal on Smart Sensing and Intelligent Systems, 14*(1), 1–9.
43. Raj, A., & Mandal, D. (2022). Comparative analysis of the 2×2 patch antenna array with modified fractal antenna for 5G communication and wireless applications. In *2022 IEEE Microwaves, Antennas, and Propagation Conference (MAPCON)* (pp. 670–675). IEEE.

44. Dhubhakarya, D. C., Srivastava, D. K., & Tiwari, M. K. (2022). Design and analysis of tri-band triangular shape fractal antenna for wireless applications. DOI: 10.21203/rs.3.rs-1926169/v1.
45. Raj, A., Dhubkarya, D. C., Srivastava, D. K., & Mandal, D. (2023). Design and analysis of square shape slot cut high-gain Sierpinski carpet fractal antenna for wireless applications. *Microwave and Optical Technology Letters, 65*(8), 2337–2343.
46. Raj, A., & Mandal, D. (2023). Design of slot cut modified rectangular shape and square shape antenna for L band and 5G applications. In *2023 First International Conference on Microwave, Antenna and Communication (MAC)* (pp. 1–4). IEEE.
47. Raj, A., & Mandal, D. (2023). Design and fabrication of high-gain array antenna for 5G communication and wireless applications. In *International Conference on Communication, Devices and Computing* (pp. 379–389). Singapore: Springer Nature Singapore.
48. Raj, A., & Mandal, D. (2024). Design and implementation of hybrid-fed array antennas for Sub-6 GHz and 5G mm-wave communication and wireless applications. *International Journal of Microwave and Wireless Technologies*, 1–20.
49. Raj, A., & Mandal, D. (2024). Design and experimental analysis of fractal antennas with ground-defected structure for expected 6G and 5G mm-wave communication and wireless applications. *Transactions on Emerging Telecommunications Technologies, 35*(1), e4900.

CHAPTER 3

Wearable Antenna for Biomedical Applications: A Comprehensive Design and Analysis Study

BHAWNA KHOKHER,[1,*] V. K. SHARMA,[1] SANJEEV SHARMA,[2] NIRAJ KUMAR,[3] VINOD KUMAR SINGH,[3] and KORHAN CENGIZ[4]

[1]*Department of Electronics and Communication, Bhagwant University, Ajmer, Rajasthan, India*

[2]*Department of Electronics and Communication, New Horizon College of Engineering, Bangalore, Karnataka, India*

[3]*Department of Electronics and Communication, Chhatrapati Shahu Ji Maharaj University, Kanpur, Uttar Pradesh, India*

[4]*Department of Computer Engineering, Istinye University, Istanbul, Türkiye*

*Corresponding author

ABSTRACT

Wearable textile antennas have become increasingly important in healthcare applications driven by miniaturization trends in wireless technologies. These antennas are gaining prominence in various communication scenarios, including public safety, mobile computing, tracking, and navigation. Design requirements for wearable antennas emphasize flat structures and flexible materials. The choice of materials significantly impacts antenna characteristics, with substrate permittivity and thickness particularly influencing the bandwidth and efficiency of microstrip antennas. Low-profile, lightweight designs are essential for wearable

Wireless Technology: Advances and Innovations for Smart IoT Applications. Rajesh Kumar Dwivedi, Vinod Kumar Singh, and Raghav Dwivedi (Eds.)

DOI: 10.1201/9781998511495-3

applications. To ensure reliable communication between wearable devices and receivers, antennas must be optimized for maximum radiation efficiency and high gain. Electromagnetic simulation software enables thorough analysis of antenna performance parameters, including radiation patterns, impedance bandwidth, and return loss. These analyses provide crucial insights for optimizing designs for specific biomedical applications. The study comprehensively explores the design and development of wearable antennas for biomedical applications, focusing on achieving reliable wireless communication in healthcare environments. This investigation encompasses the use of wearable textile materials, miniaturization techniques, and optimization approaches to enhance performance while maintaining comfort and practicality for healthcare applications.

3.1 INTRODUCTION

Wearable textile materials have been widely employed in wireless technology to create flexible and intelligent antenna segments. Wearable antennas, with their appealing features and potential for flexible, affordable, and easily accessible wireless communication and sensor applications, have drawn a lot of interest lately [1]. These antennas must be built of flexible materials and have a low-profile design since they must adapt when fastened to various human body parts, especially in the healthcare sector [2].

Body-centric interaction is possible through the textile wearable antenna system, as mentioned in Figure 3.1, especially in medical applications, IoT-based communication, defense applications, and trending technology in the wearable field. These devices need to be able to operate with the least degree of deterioration when near a person's body [3]. The antenna design is challenging due to these criteria, predominantly because of their small size, the effects of structural deformation, and connection to the body, as well as the complexity and accuracy of fabrication [4]. These difficulties have emerged in current scientific community research in this area. The majority of these issues are related to body-worn implementation, despite minor variations in severity depending on applications. Here are some difficulties in the manufacture and measurement parameters of textile wearable antennas, such as substrate material type, dielectric constant, thickness of the substrate, conductive materials, and the bending effect of the antenna and SAR.

In biomedical engineering, wearable antennas have a variety of uses, including medical diagnosis, therapy, and healthcare monitoring. To enable wireless communication between the wearable device and the outside world, these antennas are created to be incorporated into wearable devices such as smartwatches, fitness trackers, and medical equipment [5]. The antenna must be small enough to fit in a wearable device without changing its dimensions or weight. Additionally, it needs to be built with a high level of sensitivity and signal strength. Because wearable technology is so widely used, especially for health monitoring systems, the market for wearable electronics is growing [6].

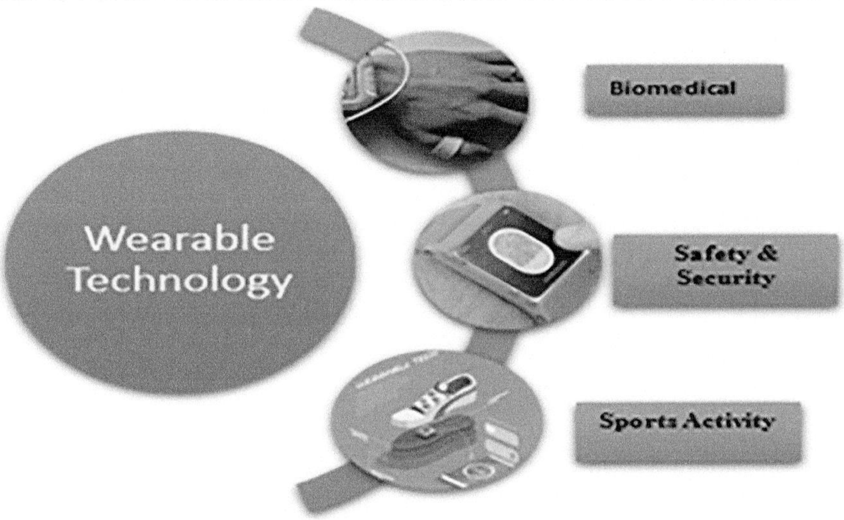

FIGURE 3.1 Trending application based on wearable technology.

The fast commercialization of IoT applications has led to the widespread adoption of wearable technology across a number of sectors, including sports and health, entertainment, and medical support. In survey straits research [7], the rapid growth of wearable technology, particularly health monitoring systems, is projected to propel the flexible electronics market to a staggering revenue of 40.37 billion by 2023. The transformative impact of smart wearables is anticipated to extend across various sectors, including business, industrial, healthcare, and sports, revolutionizing the way they operate. Wearable technology has a huge impact on the ICT sector. Additionally, it is anticipated that the worldwide enterprise wearable industry, which alone generated nearly EUR 18 billion in 2017 [7], will expand at a Compound Annual Growth Rate (CAGR) of 11.8% from 2019–2026. Figure 3.2 shows the chart of antenna market changes as well as future perspectives [8].

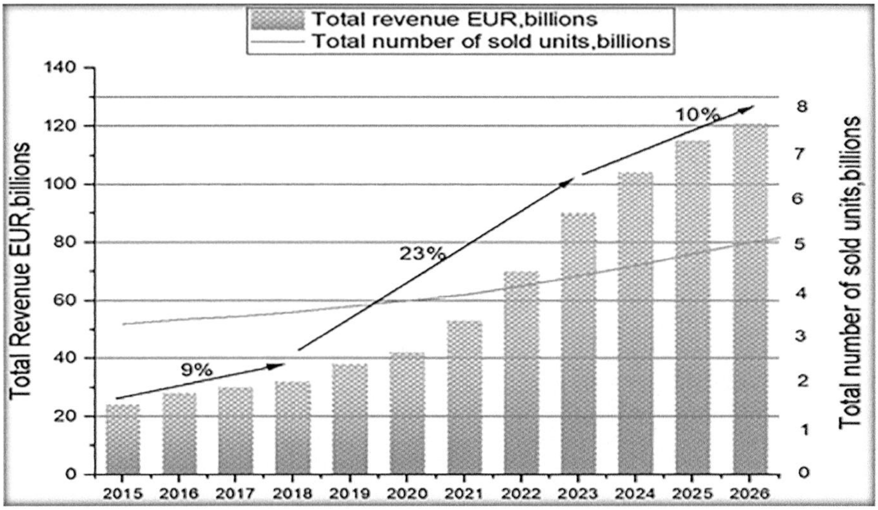

FIGURE 3.2 Growth of wearable antenna market.

Source: Reprinted with permission from Ref. [8].

The development of antenna technology during the process of miniaturization has a big influence on the production of these new wearable devices. It also provides an overview of the current research environment for wearable antennas both domestically and abroad [9]. Wearable computer devices offer a lot of potential in a range of applications, covering medical treatment, protection, and entertainment, among other uses, since they utilize wireless body area networks (WBAN) [10]. The surge in demand for flexible wireless devices is largely fueled by the increasing necessity for wearable and implantable technologies, particularly in healthcare and personal wellness. Table 3.1 outlines the diverse applications of wearable technology across various fields, showcasing its versatility and impact. In healthcare, wearable devices enable continuous monitoring of vital signs, remote patient monitoring, and early detection of health issues. Beyond healthcare, wearable technology finds applications in fitness, workplace safety, fashion, entertainment, and education, revolutionizing how we interact with technology and enhancing our daily lives. As the market for flexible wireless devices continues to grow, driven by the demand for wearable solutions, we can anticipate further advancements and expanded applications, ultimately reshaping the way we monitor health and engage with technology in everyday life.

TABLE 3.1 Wearable Technologies in Various Fields with Application

Field	Application
Biomedical field	Cancer detection, Health monitoring, Glucose monitoring Endoscopy, oximetry Wearable Device
Safety and security	Rescue- Life jackets, Helmet, trackers
Safety and security	Smart watch/intelligent shoe, Fitness band

Through wireless technology, this form of technology permits intra-body, on-body, or in-body connections with the outside world, with different areas in healthcare applications of antennas, as shown in Figure 3.3.

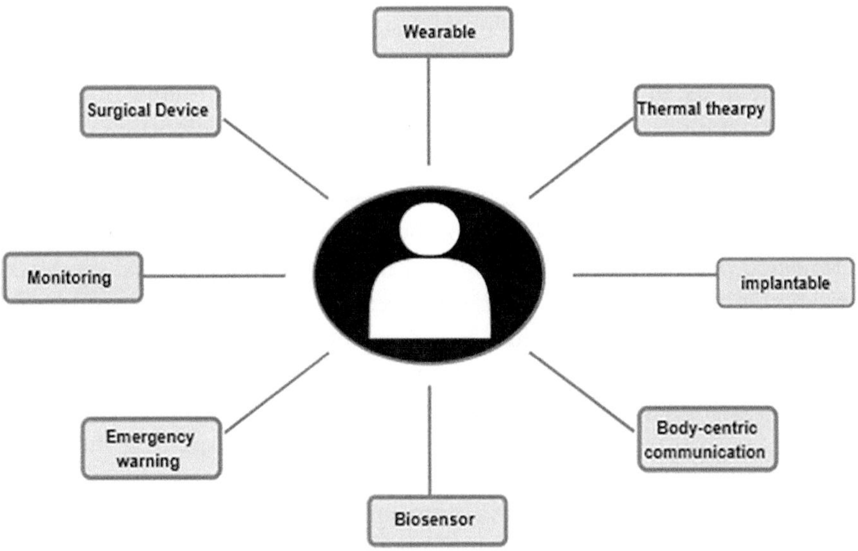

FIGURE 3.3 Various healthcare applications of wearable antennas.

Bendable printed antennas, mainly for medical applications, are required for *in vivo* applications such as intracranial sensors, continuous gait analysis, brain interfaces, drug delivery systems, health monitoring, and organ function management. The human body has curved surfaces and continuously shifting movements; therefore, devices must be physically flexible, conformal, or stretchy [12, 13] (Table 3.2).

TABLE 3.2 Mentioned Some Frequency Band Supports Wearable Antenna in Healthcare Sector

Frequency Band	Application
Ultra-high frequency	RFID tagging for medical devices, patient tracking
(ISM) bands	Wireless-body area networks (WBANs) for remote patient monitoring
2.4 GHz band (ISM)	Bluetooth connectivity for wireless sensor networks
5 GHz band	High-speed data transfer for medical imaging
60 GHz band	Ultra-high-speed data transfer for medical imaging

In Table 3.2, some frequency bands that support wearable antennas in the healthcare sector are mentioned. Achieving optimal performance with textile-based wearable antennas can pose challenges attributed to factors such as the conductivity of the materials, dielectric constant, loss tangent, substrate material thickness, and the intricacies of antenna design in bending scenarios [14]. It is significant to note that the antenna's behavior is influenced by the properties of the materials, with a planar microstrip antenna's bandwidth and efficiency being primarily determined by the type of dielectric constant and thickness of the substrate. Metamaterials, as distinct from traditional small materials like metals or plastics, are assemblages of numerous altered components that are frequently arranged in repeated patterns. Many publications have provided practical guidance for developing fully textile wearable antennas and non-textile wearable antennas for various applications based on wearable technology.

In wireless applications, the environment, transmitting power, and frequency bandwidth all impact the antenna choice and design. The antenna's performance is affected by the substrate's properties, the material used, and its fabrication technique [15]. Designing wearable antennas is challenging due to these criteria, especially when considering their small size, structural deformation effects, and attachment to the body, as well as the complexity and accuracy of fabrication [16]. These difficulties have emerged in current scientific community research in this area. Although the severity of these problems varies slightly depending on the applications, most of them are caused by body-worn implementation.

The chapter discusses the significance of materials, substrates, and design used to create flexible antennas, research trends in the conductive materials used, and their numerous applications in this context. Prospects and issues with research on flexible antennas are also highlighted. In any wireless application, the antenna design and selection are influenced by signal intensity, surroundings, and frequency range. The specifications of the substrate, the manufacturing procedure, and the antenna's performance can all impact its effectiveness and the material chosen. The advancement of conductive material research, substrates, different fabrication techniques for flexible antennas, and their numerous applications are all covered in this study. Additionally, the study highlights the potential opportunities and challenges of flexible antennas.

3.2 STUDY ANALYSIS OF WEARABLE ANTENNA

The wearable antenna has attracted the attention of many academics and has the capacity to be very significant in on-body intent communication. Since these antennas are made from fabric, wearable textiles, and flexible materials, their properties have been altered. The complete flow chart of the specifications of the substrate, the manufacturing procedure, and the antenna's performance can all have an impact on its performance, as mentioned in Figure 3.4.

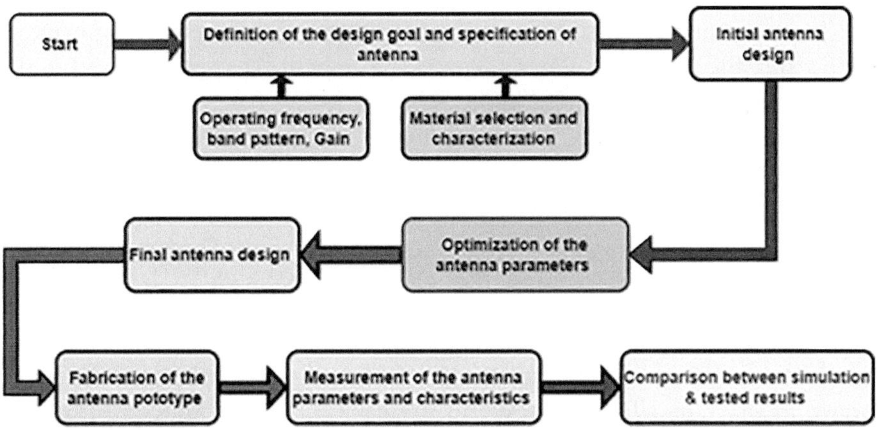

FIGURE 3.4 Simplified block diagram of an antenna design workflow.

These antennas operate close to human bodies, making their design more difficult. Additionally, these antennae must be user-friendly, strong, inexpensive, lightweight, agreeable, and maintenance-free for applicability. Materials have been chosen with considerable focus for design purposes in order to provide an acceptable amount of protection against electromagnetic radiation, minimal impact from different environmental conditions (like rain, snow, ice, etc.), and an appropriate amount of mechanical deformation (bending, twisting, and wrapping). For using fabric materials, these textiles must be properly characterized. However, using flexible non-fabric polymer-based materials like Kapton, PET, and PEN substrates has proven reliable dielectric properties [17].

3.2.1 WEARABLE ANTENNA MATERIALS

The material that is lightweight, flexible, and able to tolerate bending and straining is essential for wearable antennas. Textile materials, such as nylon, polyester, and spandex, are commonly used for flexible antennas because of their elasticity and light weight. The conductive channels in the antenna are made using conductive fabrics like copper- or silver-coated rayon [18, 19]. Elastomers are flexible, rubber-like materials that can be stretched and bent without losing their shape. In [20], liquid metal such as gallium can be printed or painted to create a conductive path onto flexible materials. These alloys have the added advantage of being self-healing and able to endure repeated bending and stretching. Conductive polymers [21] are plastics that have been modified to conduct electricity. To create conductive routes for the antenna, they can be coated or printed onto a flexible substrate. Wearable antennas can utilize graphene, a highly conductive and flexible material, as its conductor.

3.2.2 FLEXIBLE ANTENNAS CONDUCTIVE MATERIALS

To ensure high gain, efficiency, and bandwidth, conductivity is required. Nanoparticle (NP) inks, like copper and then silver, are regularly used in the construction of bendable antennas because of their excellent electrical conductivity [22, 23]. Most materials used for clothing are conductive and nonconductive textiles, as shown in Figure 3.5. Non-conductive textiles, including cotton, polyamide, jeans, Quartzel fabric, silk, and leather, are frequently used as wearable substrate materials to make wearable antennas lighter and more effective [24]. Conductive textiles (e-textiles) are innovative materials created by mixing polymer threads or conductive metals with existing fabrics. These e-textiles combine typical fabric properties with the added functionality of conductivity, including taffeta, zelt, flectron, and pure copper [8]. They make a significant contribution to the field of wearable technology. Because silver nanoparticle ink forms oxides at a lower rate than copper nanoparticle ink, it offers an advantage [33]. In addition to nanoparticles, electro-textile components like nonwoven conductive fabrics (NWCFs), Ni/Ag plating, and Flectron (copper-coated nylon fabric) are frequently used in flexible antennas [34]. An earlier article analyzed several textile and non-textile conductive materials for use in the fabrication of flexible antennas, such as copper adhesive, aluminum foil, and clad in copper [4–6]. Polyaniline [3], polypyrrole [25], and poly(3,4-ethylene dioxythiophene) polystyrene sulfonate [27] are examples of conductive polymers. Carbon nanotubes, graphene (paper, nano flakes ink, oxide ink), carbon nanoparticles, and carbon nanotubes (CNT) were used to increase the conductivity of conductive polymers, as shown in Figure 3.6 [29–33, 35]. Doping is a technique used by stretchable conductive materials to enhance conductivity so that they can withstand mechanical strain and deformation without compromising antenna performance [25]. Silicon with silver nanowire implants and fluorine rubber [31] are examples of liquid metals in a stretchable substrate. In Table 3.3, flexible materials for conductivity are mentioned.

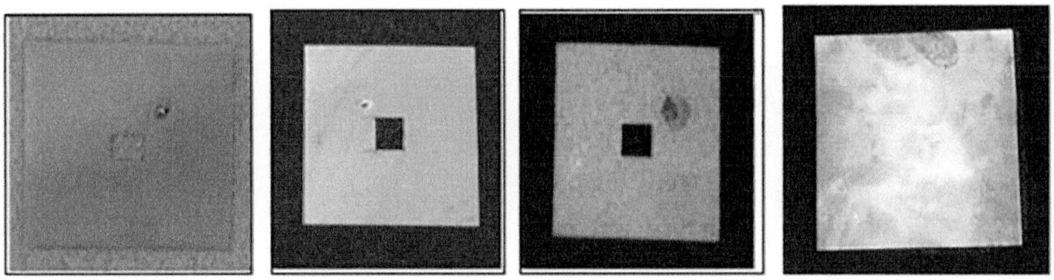

(a) (b) (c) (d)

FIGURE 3.5 Conductive materials for antenna (a) Electro textile patch (b)Patch etched in copper on thin ultra flexible polyimide substrate (c) Screen printed textile patch (d) Copper-foil on a fabric material.

Source: Reprinted with permission from [34]. © 2015 IEEE antennas and Propagation Magazine.

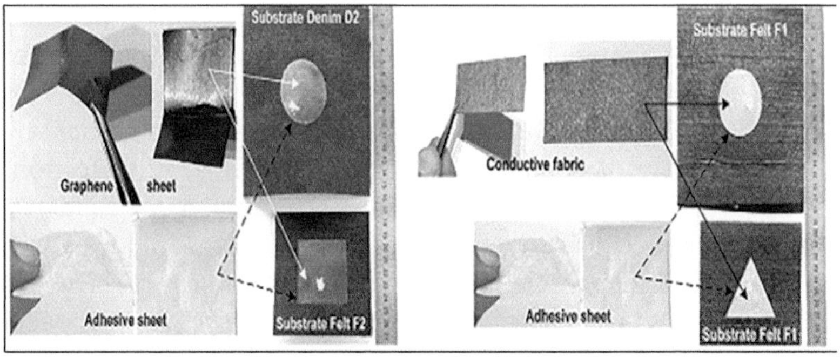

(a) (b)

FIGURE 3.6 (a) Graphene sheet as conductive materials and (b) conductive fabric on substrate.

Source: Reprinted with permission from [35]. © Electronics 2021.

TABLE 3.3 Flexible Conductive Materials with Conductivity

Material	Conductive Material Type	Conductivity σ (s/m)	References
Nano particles	Ag nano particles	2.137×10^{7}	[28, 29]
	Cu nano particles	1×10^{7}	
Conductive polymers	Polyaniline (PANI)	5	[29, 30]
	Polypyrole (PPy)	40–200	
Additive conductive polymers	C nanotube	4000–7000	[31, 35]
	Ag NW/PDMS	8130	
Graphene-based materials	Nano flake	6×10^{5}	[34, 36]
	Paper	4.2×10^{3}	
Additive conductive polymers	C nanotube	4000–7000	[31, 35]
	Ag NW/ PDMS	8130	
Graphene-based materials	Nano flakes	6×10^{5}	[34, 36]
	Paper	4.2×10^{3}	
Liquid metals	Eutectic Gain	3.4×10^{3}	[35]

3.2.3 SUBSTRATES

Highly conductive substrate materials are necessary for the efficient absorption and transmission of electromagnetic (EM) radiation.

Highly conductive fibers contained in conductive polymers, metallic inks, fabric, and polydimethylsiloxane (PDMS) contain copper and other conductive components [29–34]. The most recent types of substrates for wearable antennas are Kapton, PET, liquid crystal polymer, organic material, ferrite magnetic materials, textiles, and paper due to their distinct characteristics. Although widely used, flexible substrates like felt, fleece, silk, and Cardura exhibit limited permittivity and are lossy; yet, their effectiveness improves with larger antennas or when near the human body.

3.2.4 NORMAL TEXTILE FABRIC DIELECTRIC PROPERTIES

The textile/flexible antenna can be produced and covered with a variety of resources, including metals, inks, and textiles, depending on its intended application and specifications. Embedded polymers are tabulated in Table 3.4. For illustration, in Table 3.4, fabrics are suitable for body-worn antennas.

TABLE 3.4 Performance Parameter of Substrate Materials

Non-Conductive Substrate Materials	**Permittivity** $\in_r$	**Loss Tangent** $\tan\delta$
Felt	1.22	0.016
Cordura	1.90	0.0098
Cotton	1.60	0.0400
Tween	1.69	0.0084
Silk	1.75	0.012
Jeans	1.7	0.025
100% polyester	1.90	0.0045
Quartzel Fabric	1.95	0.0004
Leather	1.8–2.95	0.049–0.16

The environment, transmission capacity, and variable frequency have an impact on the antenna configuration and choice for each wireless application. The characteristics of the substrate, the manufacturing method, and the material used are all related to how effectively the antenna operates. The advancement of conductive material research, substrates, different fabrication techniques for flexible antennas, and their numerous applications are all covered in this study. Additionally, the study highlights the potential opportunities and challenges of flexible antennas. In Table

3.5, a summary of different textile materials' performance for biomedical applications is presented.

TABLE 3.5 Summary of Different Textile Materials Performance for Biomedical Application

Substrate	Frequency Band (GHz)	Thickness (h)	Size (mm)	$\in_r$	Application	References
Jeans	2.2–17	1	45x60	1.68	WBAN	[35]
Denim	2.45 and 5.8	0.05	46x16	1.54	Wearable and ISM	[36]
Denim	2.45 and5.8	25	42x13	1.54	Wearable devices	[38]
Jean	3–8, 9–13	1.2	28x43	1.7	Health monitoring	[39]
Polyester Fabric	1.19–4.05	0.08	45x60	2.193	Microwave medical imaging	[16]
Felt	3.1–11.3	0.7	30x 30	1.2	Wearable devices	[16]
Denim	2.61–10.1, 2.73–10.1	0.6	90x90	1.43	WBAN	[40]
denim	13.08	1	60x60	1.7	Medical Monitoring	[41]

In [41], a design was created especially for use in medical applications employing textile material as a substrate. According to modeling and test results, the proposed antenna design using substrate material is dynamic, compact in size, with a bandwidth of 13.08 GHz. To demonstrate the value of the currently proposed design in terms of bandwidth, radiation pattern, return loss, gain, and efficiency, it may monitor, inform, and demand attention in the event of a medical emergency. The proposed design's evaluation across key metrics like bandwidth, radiation pattern, return loss, gain, and efficiency highlights its capability to support reliable wireless communication during medical emergencies. This ensures timely monitoring and alerts, which are crucial for swift intervention and improved patient outcomes. By demonstrating superior performance in these areas, the design underscores its value in enhancing emergency medical response and patient care. In addition to its technical prowess, the design's focus on real-world applicability ensures seamless integration into medical monitoring systems, providing healthcare professionals with vital data when it matters most. Its efficiency in transmitting critical information in emergency situations enhances patient safety and facilitates swift decision-making (Figure 3.7).

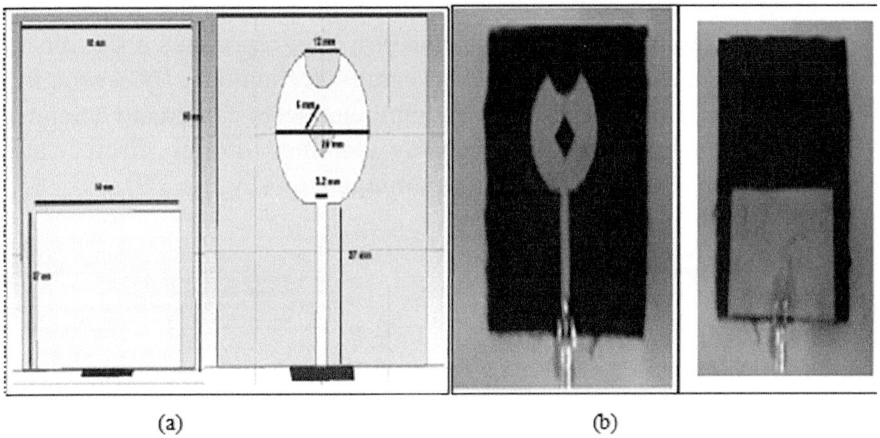

FIGURE 3.7 (a) Design part of textile antenna and (b) fabrication model on jeans substrate.

Source: Reprinted with permission from [41]. © 2015 Open Engineering.

In [14], we present compact, lightweight, low-profile textile antennas for applications in wireless sensors, as well as research into the effect of fabric materials on antennas. A stepwise process for developing, fabricating, and measuring tiny clothing-supported antennas for use in wireless devices is described. In Figure 3.8, there is a different type of fabric material to support a wearable antenna with a different substrate.

FIGURE 3.8 Flexible antenna samples with a material from denim, cotton, polyamide-elastane, and polyester.

Source: Reprinted with permission from [14]. © 2018 Sensors.

The development of flexible antenna samples using materials such as denim, cotton, polyamide-elastane, and polyester represents a significant advancement in antenna design and wearable technology. By leveraging these versatile materials, researchers, and engineers can create antennas that seamlessly integrate into everyday clothing, offering discreet and unobtrusive solutions for wireless communication (Figure 3.9).

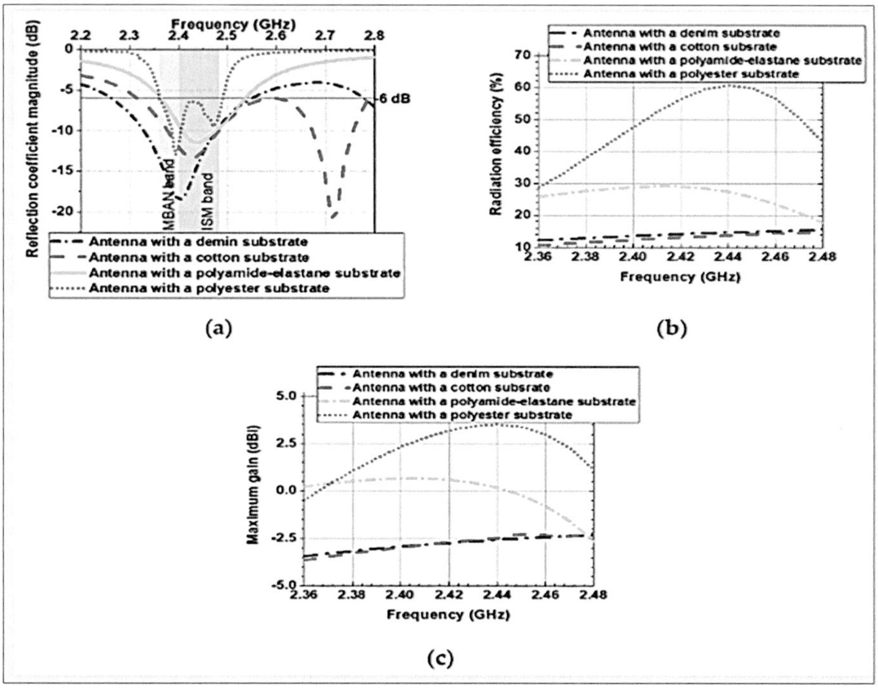

FIGURE 3.9 Simulated results with different types of substrate (a) Reflection coefficient vs frequency (b) Radiation efficiency vs frequency (c) Maximum gain vs frequency.

Source: Reprinted with permission from [16]. © 2018 Sensors.

3.3 WEARABLE ANTENNA- SPECIFIC ABSORPTION RATE (SAR) FACTOR

The rate of RF energy absorption by the human body from the source is measured by SAR. The International Commission on Non-Ionizing Radiation Protection (ICNIRP) has prescribed SAR for wireless devices like mobile phones. The Department of Telecommunication has approved the permissible SAR value limits for mobile phones to 1.6 W/kg measured over 1 g of tissue. The relationship below can be used to numerically express in Equation 3.1:

$$SAR = \frac{\sigma E^2}{\rho} \tag{3.1}$$

where σ is electrical conductivity of the tissue(S/m), E is root-mean square electric field (V/m) and ρ is the tissue's density (kg/m^3) [17].

3.3.1 SPECIFIC ABSORPTION RATE (SAR) ANALYSIS

By 1990, wireless communication started to expand significantly, at which point a SAR was required. Due to increasing concerns about the health repercussions of radiation and worldwide regulatory regulations, antenna designers, engineers, and scientists have long focused on the radiation received by tissue in the human body. The performance of the antenna when it is connected to a human body is likewise affected by human tissues.

The highest safety limits specified by the legal regulatory authority, such as IEEE C95.1–1999, for any 1 g of tissue, is less than or equal to 1.6 W/kg, and for any 10 g of tissue, is less than or equal to 2 W/kg, as determined by the ICNIRP. In [16], an Ultra-Wideband (UWB) textile antenna for an image processing system in healthcare was proposed. A low-profile antenna with a substrate of polyester fabric and conductive material copper taffeta fits the needs of wearable applications. According to [17], the measured data indicates that the antenna prototype may achieve a gain of 2.9 dBi and a bandwidth of 109% from 1.198 GHz–4.055 GHz. Also, the on-phantom tests demonstrate that the proximity of human beings to an antenna has no impact on its operational bandwidth and realized gain.

In [18], a multiple application band textile antenna for WiMAX applications, such as medical and military, was proposed. UWB applications operate at 4.62 GHz–10.74 GHz with a resonance frequency of 4.2 GHz–3.65 GHz. This study integrates a single C-shape etching slot (CSES) into a rectangular patch antenna. By integrating microstrip line feed technology with a textile substrate wearable antenna (a pair of jeans), the bending impact up to 750 and the radiation effect on human tissue as SAR are explored using computer simulation technology (CST) methods. In Table 3.6, textile antennas for numerous band applications with SAR measurement are presented.

TABLE 3.6 Comparison of Textile Wearable Antenna with SAR (W/kg)

Substrate	Antenna Size (mm)	Operational Frequency (GHz)	Design of Antenna	SAR 1 gm (W/Kg)	SAR 10 gm (W/Kg)	Application	References
Polyester Fabric	17×50	1.198–4.055	Parallel slots triangular antenna	-	0.0014	UWB	[16]
Jeans	25x25	3.4–4.3, 4.7–8.4, 10.3–14.1	Tuning fork shape antenna	Less than 1.6	Less than 2	UWB	[17]
Jeans	18×19	3.09–3.94, 4.23–5.65	C-Shape etching slot antenna	-	0.352	Wi-MAX, 5G lower band	[18]
Polyester Fabric	50×50	3.37	E-Shape patch antenna	-	-	Wi-MAX	[19]
Thin planner textile	53.6×25.5	2.45	Dual port antenna	0.12	-	ISM	[20]
Cotton	51×45	2.45	Rectangular patch antenna	-	0.486	ISM	[21]
Felt	100×100	2.45, 5.8	Circular patch antenna	0.042, 0.09	-	ISM	[22]
Polyester felt	70×70	2.45	T-Shape slot inside rectangular patch	-	0.23	ISM	[23]
Denim fabric	72×78	2.45	3-slots and 2-stubs patch antenna	-	0.22	ISM	[24]
Leather	50×50	1.65, 2.45, 3.4, 5.6	CPW Ring antenna	1.15, 1.38, 1.39, 1.5	0.683, 0.875, 1.19, 1.3	BAN	[3]

3.3.2 SAR REDUCTION TECHNIQUES

Many techniques that aim to reduce the SAR of off-body emitting antennas focus on altering the ground plane [25]. Using a periodic conductive structure, which is used to filter electromagnetic waves within frequency bands, is similar to EBG structures. The EBG deflects radiation that the planar antenna would have directed directly at the body, away from the body. The SAR levels in various body regions are checked. The maximum SAR level was 46.9 W/kg without approaches, and the lowest SAR values were 0.016 W/kg with techniques applied [26]. EBG, AMC, and metamaterials techniques perform the best in terms of lowering the SAR level [27, 28]. Using EBG [29] structural geometry at 2.4 GHz, SAR showed significant changes, with the EBG structure showing 5.41 W/kg and with EBG having 0.054 W/kg. Implanted devices expose directly to tissues when doing SAR analysis on human bodies. In Table 3.7, a comparison of the performance of SAR reduction methods is shown.

TABLE 3.7 Comparison of the Performance of SAR Reduction Methods

Operating Frequency	SAR Reduction Techniques	SAR Without Technique (W/kg)	SAR (1/10) g	SAR with Reduction Techniques (W/kg) (1/10) g	References
2.45	HIS	7.51/8.54	1	0.0257/0.0345	[25]
5.8	AMC	46.9	1	Less than 1.6	[26]
2.45	Planner antenna	-------	1	0.919/0.5176	[27]
UWB	Metamaterial	6.27	10	0.067	[28]
2.45	FSS array	5.197	1	0.164	[29]
3.5/5.8	AMC	----------	1	0.919/0.517	[30]
2.45	EBG	7.23/4.18	1	0.972/0.668	[31]

3.4 BENDING EFFECTS OF TEXTILE ANTENNA

Wearable antennas, which are antennas meant to be incorporated into clothes or worn on the body, can be affected by bending in a variety of ways. Twisting distortion makes wearable antenna performance crucial because bending can affect the structural characteristics of the antenna in several ways, including altered radiation patterns, resonance frequency shifts, mismatched impedance, signal attenuations, mechanical stress, and so forth. In [42], the author has proposed the textile patch antenna to allow wireless communications with sufficient bandwidth (Figure 3.10).

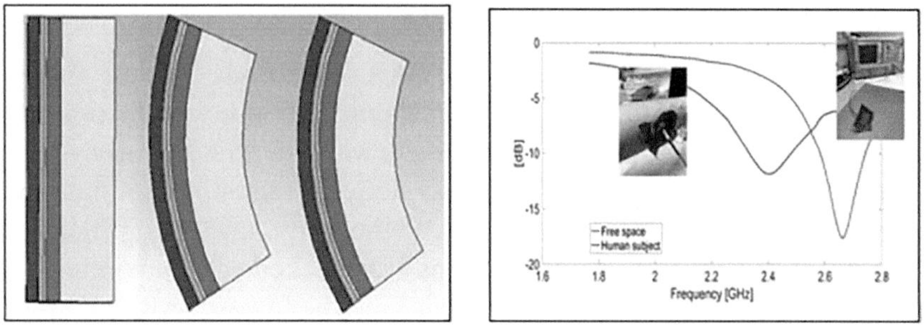

(a) (b)

FIGURE 3.10 (a) Bending effect on human tissue and (b) Comparison between return loss measurements performed into the free space and in contact with the human body.

Source: Reprinted with permission from [43]. © 2018 IEEE Transactions on Antennas and Propagation.

In [43], research is conducted into wearable patch antenna bending across cylindrical surfaces. Using the design and testing of patch antennas bent at various angles, significant frequency fluctuations and radiation pattern deviations have been reviewed. The frequency shift diagram concerning bending radius has been designed for multiple wearable applications. Analysis has been done to determine how resonant frequency shifting affects alterations in the E and H planes. The main purpose of the project is to create helpful design-oriented charts and provide wearable electronics designers with an enhanced estimation of how antenna curving affects their products (Figures 3.11–3.13).

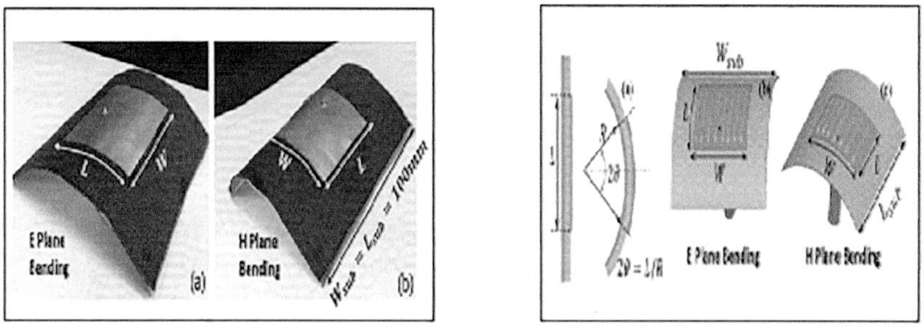

(a) (b)

FIGURE 3.11 E plane and H plane bending (a) Printed and (b) Design of proposed bending.

Source: Reprinted with permission from [44]. © 2018 IEEE Access.

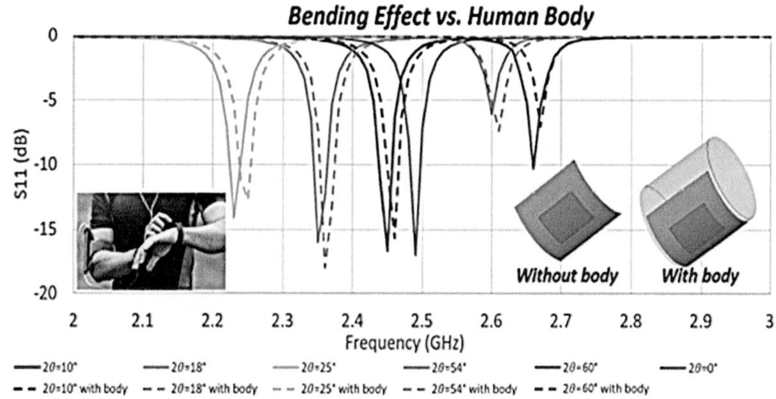

FIGURE 3.12 Recover S11 for the E plane bent patch antenna with and without the cylindrical phantom lossy tissue.

Source: Reprinted with permission from [44]. © 2018 IEEE Access.

In [44], the proposal and demonstration of a miniature textile antenna for ISM band purposes at 2.4 GHz are presented. The suggested antenna creates an inverted E-shaped design by properly loading a rectangular slot or notch with an added strip line. The structure is simple, compact, and easy to construct, using fabric as the material. An ordinary antenna is 75% larger than this one. The antenna's resistance to deformation is demonstrated by how well it performs when bent.

The bending effect of flexible antennas refers to how the antenna's performance is influenced by its bending or curvature. As flexible antennas are designed to bend or conform to various shapes and surfaces, their electromagnetic properties may be altered when subjected to bending stresses. When a flexible antenna bends, it can experience changes in its resonance frequency, impedance matching, radiation pattern, and efficiency. These changes depend on factors such as the antenna's design, materials used, bending radius, and the degree of bending. In some cases, bending can cause deformation or distortion of the antenna structure, leading to impedance mismatches, increased losses, and reduced radiation efficiency. Additionally, bending may also affect the mechanical integrity of the antenna, potentially leading to fatigue and failure over time. To mitigate the bending effect, designers of flexible antennas employ various strategies, such as using flexible and durable materials, optimizing the antenna's mechanical design to withstand bending stresses, and incorporating adaptive tuning mechanisms to maintain performance despite changes in shape (Figures 3.14 and 3.15).

FIGURE 3.13 Bending effect of the proposed antenna at various radii in the X-and Y-direction.
Source: Reprinted with permission from [35]. © 2021 Electronics.

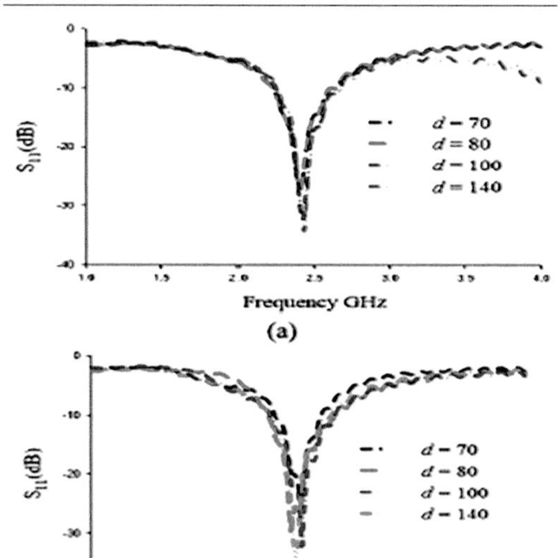

FIGURE 3.14 Measured S_{11} (dB) at various bending deformation along (a) x-axis (b) y-axis.
Source: Reprinted with permission from [35]. © 2021 Electronics.

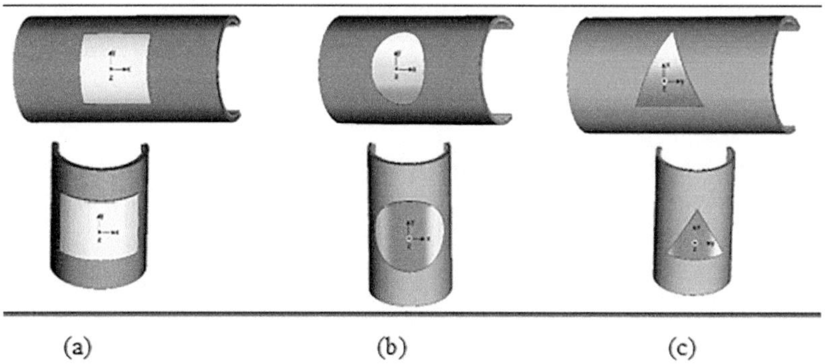

FIGURE 3.15 Bending conditions for (a) the rectangular patch, (b) the circular patch, and (c) the triangular patch antennas.

Source: Reprinted with permission from [35]. © 2021 Electronics.

Understanding and characterizing the bending effect of flexible antennas is crucial for ensuring their reliable operation in real-world applications, particularly in environments where bending or conformability is required, such as wearable devices, flexible displays, and Internet of Things (IoT) sensors. To address the bending effect, designers may employ advanced materials with enhanced flexibility and mechanical properties, such as flexible polymers, elastomers, or composite materials. Additionally, innovative antenna designs, such as meandered or serpentine structures, can help mitigate the impact of bending by distributing mechanical stresses more evenly across the antenna's surface (Figure 3.16).

3.5 CONCLUSION

Wearable textile antennas have become increasingly important in modern healthcare. The fabrication of microstrip antenna segments makes considerable use of fabric resources because of their affordability, versatility, lightweight, and conformal design, which is in line with current miniaturization developments. The purpose of this review work is to provide a general understanding of wearable/textile antennas and their applications in healthcare segments. The critical challenge in designing wearable antennas is the material selection, the design and dimensions of the antenna, and on/off body interaction. In wearable/textile antennas, material selection depends on properties such as conductivity, permittivity, and

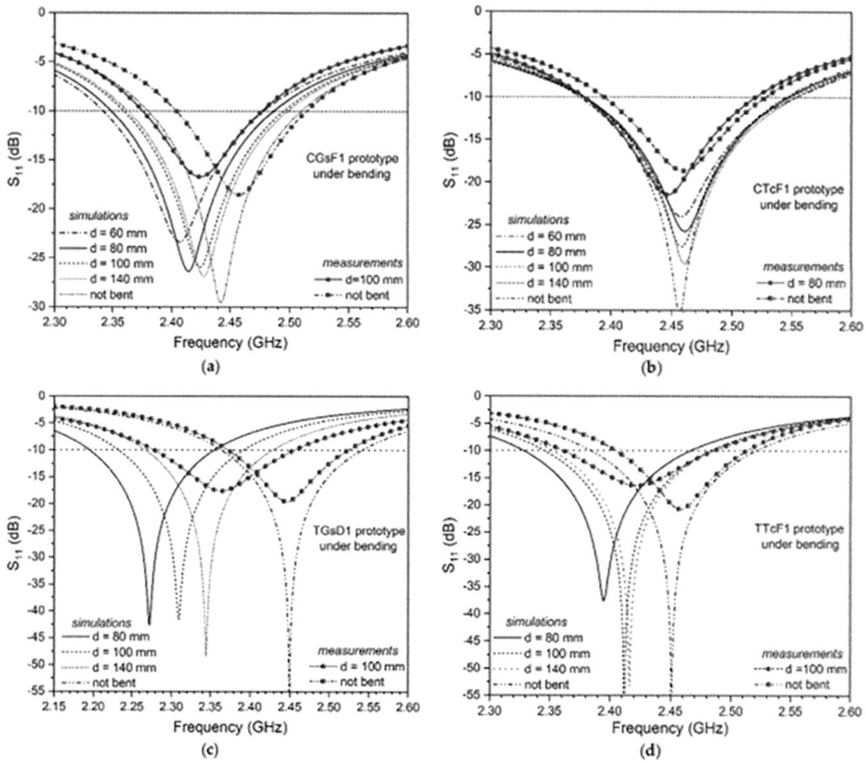

FIGURE 3.16 Measured and simulated magnitude of S11 versus frequency for (a) the circular graphene sheet patch CGsF1, (b)circular conducting fabric patch CTcF1, (c) triangular graphene sheet patch TGsD1, and (d) triangular conductive patchTTcF1 antenna prototype.

Source: Reprinted with permission from [35]. © 2021 Electronics.

loss tangent, which play an important role in designing an antenna. While the substrate is the most important component of the wearable antenna, the material provides better performance and range. The review work provides a summary of the different textile materials' performance for biomedical applications in Table 3.2. Furthermore, testing with the human body is required to achieve an antenna design that meets the specifications of wearable antennas. As a result, it was tested on human tissue (1 g/10 g) with absorption rate performance. Furthermore, as Table 3.4 illustrates, wearable antenna performance might differ among various human body

TABLE 3.8 Performance of the Wearable Antennas After Bending

Size	Substrate	Bending	Radius	Frequency Range Before	Frequency Range After	Bandwidth	References
44.6×44.6 mm²	Denim material	Flat, Male, Female	Male- 5, Female- 4.40	2.4 GHz	2.4 GHz	-	[42]
100×100 mm²	Cotton material	Flat, E-plane, H-plane	35, 60, 105, 160, 250, Flat	2.2 GHz, 2.3 GHz, 2.4 GHz, 2.4 GHz, 2.6 GHz, 2.6 GHz	2.4 GHz, 2.5 GHz, 2.5 GHz, 3.8 GHz, 3.9 GHz, 3.9 GHz	-	[43]
40×60 mm²	Denim material	Vertical, Horizontal, Flat	70, 80, 100, 140	2.4 GHz	Vertical- 2.23–2.57, 2.24–2.58, 2.23–2.58, 2.26–2.62; Horizontal- 2.28–2.57, 2.2;4–2.57, 2.23–2.59, 2.25–2.6; Flat- 2.23–2.59	Ver-340, 340, 350, 360; Hor- 290, 330, 360, 350; Flat-360 MHz	[44]
81×106 mm²	Thick denim	Vertical, Horizontal	140, 100, 80, 60	2.428 GHz	2.461 GHz	131 MHz–197 MHz	
70×103 mm²	Thin denim	Vertical, Horizontal	140, 100, 80, 61	2.425 GHz	2.459 GHz	131 MHz–197 MHz	
101×118 mm²	Thick felt	Vertical, Horizontal	140, 100, 80, 62	2.415 GHz	2.460 GHz	109 MHz–178 MHz	[35]
90×113 mm²	Thin felt	Vertical, Horizontal	140, 100, 80, 63	2.407 GHz	2.458 GHz	109 MHz–178 MHz	

types, affecting factors such as radiation pattern, operating frequency, bandwidth, and efficiency. It is therefore imperative to introduce a variety of SAR reduction approaches, which could guarantee wearable antennas' exceptional performance and stability. In conclusion, to provide a high level of isolation between the antenna and the human body and to improve antenna performance, techniques such as EBG, PEC, ferrite sheets, and so on, are described in Table 3.7. In view of the above-mentioned facts, wearable antenna performance depends on the selection of materials, design parameters, and the reduction technique of the SAR for the desired application.

KEYWORDS

- **bending**
- **flexible antenna**
- **microstrip antenna**
- **SAR**
- **textile antennas**
- **wearable antenna**
- **wearable devices**
- **wireless devices**

REFERENCES

1. Kaur, H., & Chawla, P. (2022). Performance analysis of novel wearable textile antenna design for medical and wireless applications. *Wireless Personal Communications, 124*(2), 1475–1491.
2. Ali, S. M., Sovuthy, C., Imran, M. A., Socheatra, S., Abbasi, Q. H., & Abidin, Z. Z. (2020). Recent advances of wearable antennas in materials, fabrication methods, designs, and their applications: State-of-the-art. *Micromachines, 11*(10), 888.
3. Salvado, R., Loss, C., Gonçalves, R., & Pinho, P. (2012). Textile materials for the design of wearable antennas: A survey. *Sensors, 12*(11), 15841–15857.
4. Paracha, K. N., Rahim, S. K. A., Soh, P. J., & Khalily, M. (2019). Wearable antennas: A review of materials, structures, and innovative features for autonomous communication and sensing. *IEEE Access, 7*, 56694–56712.

5. Lee, J., Kim, D., Ryoo, H. Y., & Shin, B. S. (2016). Sustainable wearables: Wearable technology for enhancing the quality of human life. *Sustainability, 8*(5), 466.
6. Naik, K. K., Teja, S. C. S., Sailaja, B. V., & Sri, P. A. (2020). Design of flexible parasitic element patch antenna for biomedical application. *Progress In Electromagnetics Research M, 94*, 143–153.
7. Llamas, R. T. (2019). *Enterprise Wearable Market: Information by Technology (IoT, Bluetooth, BLE), Product (Wristwear, Footwear, Eyewear), Application (Infotainment, IT), and Regional Outlook–Forecast Till 2026*. Tech. Rep. SR1195, Straits Research.
8. I-SCOOP. (2016). *Wearables Market Outlook: Drivers and New Markets*. Retrieved from: https://www.i-scoop.eu/wearables-market-outlook-2020-drivers-newmarkets/ (accessed on 30 January 2025).
9. Liu, Y., Wang, H., Zhao, W., Zhang, M., Qin, H., & Xie, Y. (2018). Flexible, stretchable sensors for wearable health monitoring: Sensing mechanisms, materials, fabrication strategies, and features. *Sensors, 18*(2), 645.
10. Singh, S., & Prasad, D. (2022). Wireless body area network (WBAN): A review of schemes and protocols. *Materials Today: Proceedings, 49*, 3488–3496.
11. Gao, W., Zhu, Y., Wang, Y., Yuan, G., & Liu, J. M. (2020). A review of flexible perovskite oxide ferroelectric films and their application. *Journal of Materiomics, 6*(1), 1–16.
12. Butt, M. A., Kazanskiy, N. L., & Khonina, S. N. (2023). Flexible and wearable sensing devices: An introduction. In *Flexible and Wearable Sensors: Materials, Technologies, and Challenges* (pp. 1–20). CRC Press.
13. Corchia, L., Monti, G., De Benedetto, E., & Tarricone, L. (2017). Wearable antennas for remote health care monitoring systems. *International Journal of Antennas and Propagation*, 1–11.
14. Atanasova, G., & Atanasov, N. (2020). Small antennas for wearable sensor networks: Impact of the electromagnetic properties of the textiles on antenna performance. *Sensors, 20*(18), 5157.
15. Choi, C., Bansal, S., Münzenrieder, N., & Subramanian, S. (2021). Fabricating and assembling acoustic metamaterials and phononic crystals. *Advanced Engineering Materials, 23*(2), 2000988.
16. Lin, X., Chen, Y., Gong, Z., Seet, B. C., Huang, L., & Lu, Y. (2020). Ultrawideband textile antenna for wearable microwave medical imaging applications. *IEEE Transactions on Antennas and Propagation, 68*(6), 4238–4249.
17. Shin, K. Y., Hong, J. Y., & Jang, J. (2011). Micropatterning of graphene sheets by inkjet printing and its wideband dipole-antenna application. *Advanced Materials, 23*(18), 2113–2118.
18. Zeng, W., Shu, L., Li, Q., Chen, S., Wang, F., & Tao, X. M. (2014). Fiber-based wearable electronics: A review of materials, fabrication, devices, and applications. *Advanced Materials, 26*(31), 5310–5336.
19. Shi, J., Liu, S., Zhang, L., Yang, B., Shu, L., Yang, Y., & Tao, X. (2020). Smart textile-integrated microelectronic systems for wearable applications. *Advanced Materials, 32*(5), 1901958.

20. Chang, H., Guo, R., Sun, Z., Wang, H., Hou, Y., Wang, Q., & Liu, J. (2018). Direct writing and repairable paper flexible electronics using nickel–liquid metal ink. *Advanced Materials Interfaces, 5*(20), 1800571.
21. Teghai, G. B., Mengistie, D. A., Malengier, B., Fante, K. A., & Van Langenhove, L. (2020). PEDOT: PSS-based conductive textiles and their applications. *Sensors, 20*(7), 1881.
22. Islam, G. N., Ali, A., & Collie, S. (2020). Textile sensors for wearable applications: A comprehensive review. *Cellulose, 27*, 6103–6131.
23. Kim, Y. S., Basir, A., Herbert, R., Kim, J., Yoo, H., & Yeo, W. H. (2019). Soft materials, stretchable mechanics, and optimized designs for body-wearable compliant antennas. *ACS Applied Materials & Interfaces, 12*(2), 3059–3067.
24. Yang, W., Cheng, X., Guo, Z., Sun, Q., Wang, J., & Wang, C. (2023). Design, fabrication, and applications of flexible RFID antennas based on printed electronic materials and technologies. *Journal of Materials Chemistry C, 11*(2), 406–425.
25. Park, M., Im, J., Shin, M., Min, Y., Park, J., Cho, H., & Kim, K. (2012). Highly stretchable electric circuits from a composite material of silver nanoparticles and elastomeric fibers. *Nature Nanotechnology, 7*(12), 803–809.
26. Sebastian, M. T., Ubic, R., & Jantunen, H. (2017). *Microwave Materials and Applications* (Vol. 2, p. 401).
27. Wong, W. S., & Salleo, A. (2009). *Flexible Electronics: Materials and Applications* (Vol. 11). Springer Science & Business Media.
28. Khan, S., Lorenzelli, L., & Dahiya, R. S. (2014). Technologies for printing sensors and electronics over large flexible substrates: A review. *IEEE Sensors Journal, 15*(6), 3164–3185.
29. Zahir, H., Wojkiewicz, J. L., Alexander, P., Kone, L., Belkacem, B., Bergheul, S., & Lasri, T. (2016). Design, fabrication, and characterization of polyaniline and multiwall carbon nanotubes composites-based patch antenna. *IET Microwaves, Antennas & Propagation, 10*(1), 88–93.
30. Subbaraman, H., Pham, D. T., Xu, X., Chen, M. Y., Hosseini, A., Lu, X., & Chen, R. T. (2013). Inkjet-printed two-dimensional phased-array antenna on a flexible substrate. *IEEE Antennas and Wireless Propagation Letters, 12*, 170–173.
31. Kuila, T., Bose, S., Khanra, P., Mishra, A. K., Kim, N. H., & Lee, J. H. (2011). Recent advances in graphene-based biosensors. *Biosensors and Bioelectronics, 26*(12), 4637–4648.
32. Zhu, S., So, J. H., Mays, R., Desai, S., Barnes, W. R., Pourdeyhimi, B., & Dickey, M. D. (2013). Ultrastretchable fibers with metallic conductivity using a liquid metal alloy core. *Advanced Functional Materials, 23*(18), 2308–2314.
33. Kirtania, S. G., Elger, A. W., & Hasan, M. R. (2020). Flexible antennas: A review. *Micromachines, 11*(9), 847.
34. 36. Nepa, P., & Rogier, H. (2015). Wearable antennas for off-body radio links at VHF and UHF bands: Challenges, the state of the art, and future trends below 1 GHz. *IEEE Antennas and Propagation Magazine, 57*(5), 30–52.
35. Kapetanakis, T. N., Nikolopoulos, C. D., Petridis, K., & Vardiambasis, I. O. (2021). Wearable textile antenna with a graphene sheet or conductive fabric patch for the 2.45 GHz band. *Electronics, 10*(21), 2571.

36. Turkmen, M., & Yalduz, H. (2018). Design and performance analysis of a flexible UWB wearable textile antenna on jeans substrate. *International Journal of Information and Electronics Engineering, 8*(2), 15–18.
37. Wang, K. H., & Li, J. (2018). Jeans textile antenna for smart wearable antenna. In *2018 12th International Symposium on Antennas, Propagation and EM Theory (ISAPE)* (pp. 1–3). IEEE.
38. Li, S. H., & Li, J. S. (2018). Smart patch wearable antenna on jeans textile for body wireless communication. In *2018 12th International Symposium on Antennas, Propagation and EM Theory (ISAPE)* (pp. 1–4). IEEE.
39. Jayabharathy, K., & Shanmuganantham, T. (2019). Design of a compact textile wideband antenna for smart clothing. In *2019 2nd International Conference on Intelligent Computing, Instrumentation and Control Technologies (ICICICT)* (Vol. 1, pp. 477–481). IEEE.
40. Di Natale, A., & Di Giampaolo, E. (2020). A reconfigurable all-textile wearable UWB antenna. *Progress In Electromagnetics Research C, 103*, 31–43.
41. Singh, N., Singh, A. K., & Singh, V. K. (2015). Design & performance of wearable ultra-wide band textile antenna for medical applications. *Open Engineering, 57*(7), 117–123.
42. Costanzo, S., & Cioffi, V. (2020). Full analysis of wearable textile sensor for biomedical applications: Preliminary validations towards a pre-clinical assessment. In *2020 14th European Conference on Antennas and Propagation (EuCAP)* (pp. 1–5).
43. Song, L., & Rahmat-Samii, Y. A. (2018). Systematic investigation of rectangular patch antenna bending effects for wearable applications. *IEEE Transactions on Antennas and Propagation, 66*(5), 2219–2228.
44. Ashyap, A. Y., Abidin, Z. Z., Dahlan, S. H., Majid, H. A., Waddah, A. M. A., Kamarudin, M. R., & Noras, J. M. (2018). Inverted E-shaped wearable textile antenna for medical applications. *IEEE Access, 6*, 35214–35222.

CHAPTER 4

Design of a Hybrid Bio-Inspired Model for Energy Harvesting Networks

PARUL AWASTHI and VISHAL AWASTHI*

Department of Electronics and Communication Engineering, UIET, CSJM University, Kanpur, Uttar Pradesh, India

*Corresponding author

ABSTRACT

Energy harvesting for maximizing battery efficiency is a pivotal task in wireless sensor networks. Sustainable sources of energy are converted into electrical energy with an energy harvesting circuit. There is a need for fundamentally efficient models that allow for low-energy operations. Energy distribution modeling is required while designing energy-collecting networks, but they are susceptible to internal and external defects, which increase energy consumption. The proposed model uses a Genetic Algorithm (GA) based on their residual energy and distance metrics. Particle Swarm Optimization (PSO) is a technique for processing clustered nodes that assists in the deployment of a fault-tolerant and energy-harvesting process. The PSO method is further enhanced by combining an Ant Colony Optimization (ACO) model and Teacher Learner Based Optimization (TLBO), which aids in value-based defect identification and mitigation activities.

4.1 INTRODUCTION

Optimizing battery use is a critical job for any wireless sensor network. A broad range of models is created to accomplish this goal, including energy harvesting, duty cycle optimization, node-to-node load balancing,

Wireless Technology: Advances and Innovations for Smart IoT Applications. Rajesh Kumar Dwivedi, Vinod Kumar Singh, and Raghav Dwivedi (Eds.)

DOI: 10.1201/9781998511495-4

distributed computing, and so on. Nevertheless, due to restricted battery capacity and physical access restrictions, WSN nodes frequently require battery replacement, reducing their long-term operational efficiency. Several energy-collecting circuits are included in wireless nodes to lessen the likelihood of battery replacement [1].

Sustainable sources of energy like solar energy, wind energy, temperature changes, mechanical disturbances, and so on are converted into electrical energy via these circuits. This converted energy is utilized to charge the wireless sensor node attached to the power source. Furthermore, as shown in Figure 4.1, some mobile systems offload computationally intensive jobs to highly energetic nodes in order to lengthen network lifespan and hence improve battery capacity.

According to the network concept, each energy harvesting node (EHN) contains a battery source and an energy harvesting circuit. This circuit is in charge of turning renewable energy sources into electrical power to charge the battery. The EHNs evaluate their battery power and offload computationally difficult tasks to the center's wireless sensor node (WSN). This method improves the offloading node's energy efficiency, providing it more time for outside charging.

Energy harvesting in wireless sensor networks necessitates the development of fundamentally efficient models that allow for low-energy operations. For example, the work in Deb, M, Roy, S. [2] presents a low-energy clustering approach that harvests solar energy. Solar electricity is harvested using components such as a solar plate, harvesting module device, and storage device. The gathered power is expressed as follows in Equation 4.1:

$$P_{harvest}(t) = P_{battery}(t+1) - P_{battery}(t) + P_{sleep}(t) \tag{4.1}$$

Where, $P_{harvest}$, $P_{battery}$, P_{active}, *and* P_{sleep} represent harvested, battery, active, and sleep power levels. The model also makes use of a customized frame structure that supports the EH model, with the following parameters kept on the frame and conveyed throughout the network: information sent between nodes, calculating the distance between nodes, cluster head selection information, cluster formation details, node identification number, rate of energy harvesting, and residual energy in the nodes are all calculated for clustering.

Energy harvesting in multimedia networks considers energy efficiency while maintaining high throughput and bundle conveyance execution. [4]

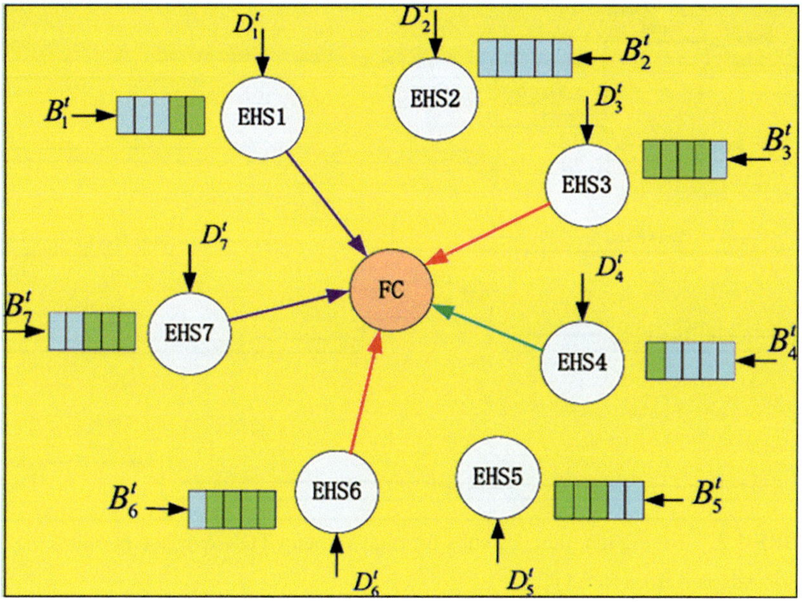

FIGURE 4.1 Wireless sensor network with energy harvesting.

Source: Adapted from Ref. [3]

Suggests the use of energy-saving pioneering directing (ESOR) to extend network lifespan for generic remote sensor organizations (WSNs). The framework reduces energy consumption by putting idle hubs to sleep and managing their waking cycles based on the remaining energy of hubs and the organization's stream rate. High-velocity time division access (TDMA)-based scheduling is used to arrange these cycles. Figure 4.2 depicts an example of this planning by displaying the rest and dynamic times of hubs based on their needs. Because of this configuration, overall energy productivity is increased by 55%, extending the network lifespan. For application-specific execution, this computation should be tested on clear organization traffic such as multimedia traffic. Moreover, the calculation's efficacy may be increased by including principles such as warm entropy awareness, as mentioned in Ref. [5].

4.2 DESIGNING ENERGY HARVESTING NETWORKS

Energy distribution modeling under various real-time network situations is required while designing energy-collecting networks. These networks

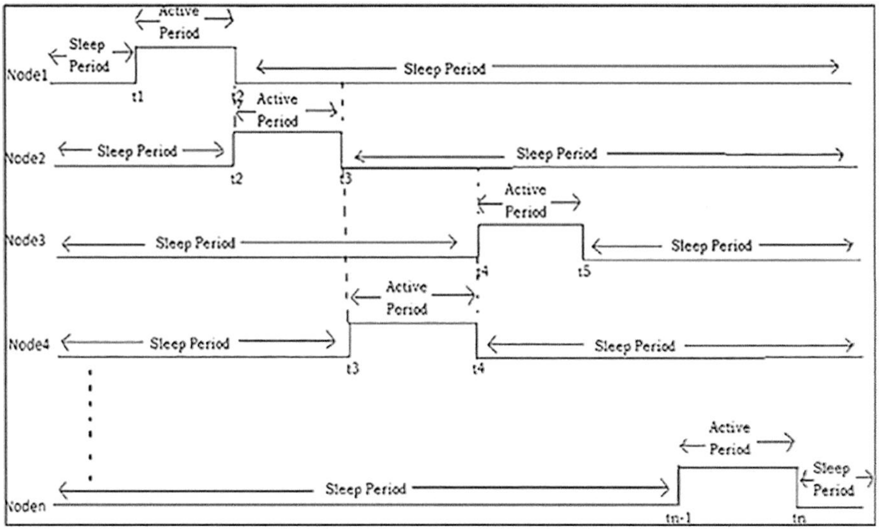

FIGURE 4.2 Timing diagrams of hubs for high velocity TDMA-based booking.
Source: Adapted from Ref. [1]

are more energy-efficient, but they are susceptible to internal and external defects, which increase the energy consumption of afflicted nodes. Because of this, the likelihood of node failure and network failure increases, lowering QoS (Quality of Service) for network deployment. Researchers have developed several fault tolerance and mitigation methods to address this issue; however, these models require substantial training datasets and real-time samples to operate efficiently. This raises the network's computational complexity, storage costs, and end-to-end processing latency, lowering its QoS performance in real-time use scenarios. To address these difficulties, this chapter suggests using fuzzy rule checks to develop a hybrid bio-inspired model for fault-tolerant energy harvesting networks.

The proposed model begins by grouping nodes using a Genetic Algorithm based on their residual energy and distance metrics (GA). Particle Swarm Optimization (PSO) is a technique for processing clustered nodes that assists in the deployment of a fault-tolerant and energy-harvesting process. The PSO method is further enhanced by combining an Ant Colony Optimization (ACO) model and Teacher Learner Based Optimization (TLBO), which aids in value-based defect identification and mitigation activities.

During network deployment, all bio-inspired models are trained once and then evaluated for each communication request. The model reassesses average QoS performance after a specific number of interactions and gradually reconfigures preferred solutions. In contrast to other state-of-the-art models, the model consumes less energy and has less complexity as a result of this gradual adjustment. Compared to earlier published energy harvesting methods, the proposed approach has 15.4% lower energy usage, 8.5% quicker communication response, 9.2% higher throughput, and 1.5% higher packet delivery ratio (PDR).

In comparison to its competitors, the proposed model showcases notable advantages in fault prediction and mitigation capabilities, rendering it highly suitable for deployment across various real-time network installations. The design of an energy recovery model tailored for Wi-Fi communication entails a comprehensive multi-domain approach encompassing several crucial components. These include node analysis, fault assessment, mitigation strategy estimation, energy transfer mechanisms between nodes, and the preservation of Quality of Service (QoS) through the application of Machine Learning Models, among other considerations. To develop an efficient energy harvesting scheme, researchers, and network designers must thoroughly investigate numerous network-specific and node-specific factors. Figure 4.3 illustrates a typical architecture for failure-aware energy recovery, leveraging insights from Cuckoo Seo Services (CSO) and integrating Deep Reinforcement Learning (DRL) techniques for retraining and energy optimization purposes. At the heart of this paradigm lies an energy harvesting block, facilitating the distribution of energy across nodes within large-scale WSNs. Building upon this foundation, a pre-processing block enhances the efficacy of energy sharing strategies by proactively identifying nodes with low energy levels, evaluating the trustworthiness of high-energy nodes, and identifying nearby nodes to optimize communication performance. This architecture embodies a proactive approach to energy management within Wi-Fi communication networks, wherein fault prediction, mitigation, and energy optimization strategies are seamlessly integrated to ensure robust and reliable network operation. By harnessing advanced computational techniques such as DRL and leveraging insights from CSO methodologies, the proposed model demonstrates a holistic approach to addressing the complex challenges associated with energy harvesting and management in wireless communication environments. Through the implementation of such innovative architectures, network stakeholders can enhance network resilience, optimize energy utilization, and ensure sustained performance across diverse operational scenarios.

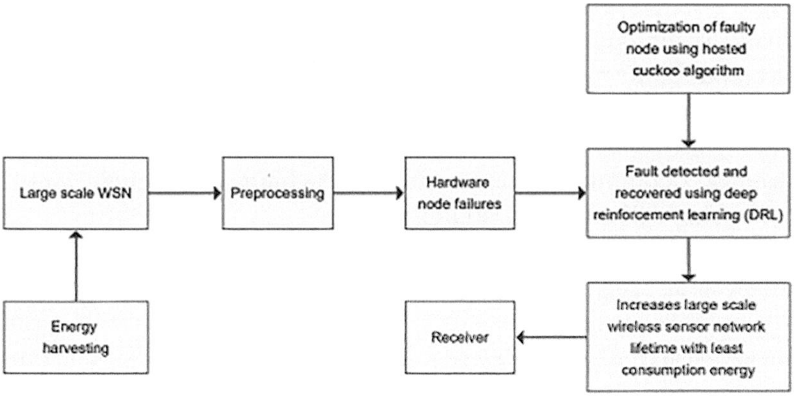

FIGURE 4.3 Design of a typical energy harvesting model for fault-tolerant WSN.

A DRL model is utilized to assess the identified nodes for hardware failure, facilitating the identification of problematic nodes and removing them from the communication process. The nodes that are selected are filtered using the fitness assessment function described in Equation 4.2, which assists in locating nodes with higher energy and lower distance measurements.

$$f_i = \left(\sqrt{\frac{(x_1 - x_2)^2 + (y_1 - y_2)^2}{E_{final} - E_{initial}}} \right)_i \tag{4.2}$$

Where, (x) represents the location of the nodes, while E_{final} and E_{initial} represent the final and initial energy of nodes involved in the communication process. Researchers offer similar methods that aid in minimizing energy usage by identifying high-trust nodes using various fitness factors. The next section of this chapter discusses such energy-efficient and fault-tolerant models, as well as their nuances, benefits, limitations, and future research goals. Several of these architectures, according to the discussion, include internal and external defects that raise the energy consumption of impacted nodes. While alternative fault-tolerant models are sophisticated, they increase the network's computational complexity, storage cost, and end-to-end processing delay, lowering QoS performance in real-time use scenarios.

4.3 LITERATURE REVIEW

The researchers propose a variety of energy harvesting and fault-tolerant models, each with distinct computational qualities that impact their deployment options. For example, the studies in Refs. [6, 7] provide RF-based harvested energy for IoT devices, a data and energy central system (DEIN), and an improved uneven cluster protocol (IUCP) to reduce node-level energy utilization during network communications. The work presented in Ref. [8] explores the integration of energy-conscious net joining techniques with Device-Selective Energy Requests (DSERs), Upper Credibility Bound (UCB), and moment channel hopping (TSCH).

Due to the intrinsic use of energy preservation mechanisms, which assist in decreasing continuous power consumption under diverse communication settings, these models demonstrate high energy efficiency. Nevertheless, the intricacy of these models limits their applicability and scalability. To address this limitation, research in Ref. [9] suggests using Energy-Efficient Task Offloading through Differential Evolution, as well as an Edge Computing Model, which assists in low-power, high-efficiency network deployments.

This design makes use of cloud computing to tackle exceedingly complex tasks while offloading low-power tasks to lower-performance nodes. As a consequence, the model has enhanced its energy efficiency and throughput capacities. Based on this technique, research in Refs. [10–12] introduces Energy Harvesting Intelligent Relay Selection Protocol (EH-IRSP), Online Energy Scheduling, and dynamic programming (DP) to assist in boosting throughput while maintaining high energy efficiency in a variety of network scenarios. Researchers also discuss models that use joint caching and user association for energy operations [13], FTF [14], capacitor management schemes [15], and Extended Hierarchical Clustering (EHC) [16], all of which aim to reduce power consumption in wireless sensor networks through redundancy estimation and removal.

Researchers suggest the Cutoff Point and Electricity Enabled Multilevel Stable Elections Protocol (TEMSEP) and Regional Deep Learning Algorithm (DDRL) for energy Recovery Virtualized Micro Cells in [17–19]. These models may blend energy levels with distance and other aspects to increase the energy efficiency of WSN installations. Extensions to these models, such as Optimal Resource Allocation, Orientation-Impartment Multiple Input Energy Efficient Networks, Multiple Showcased Actor and Critic-based Relaying Models, and Multiple Featured Actor and Critic based Relay Selection Models, are discussed in Refs. [20–22] for dynamic networks. These models show how to address energy efficiency in wireless situations by using both network and node characteristics.

To improve wireless network efficiency in real-time applications, fault-tolerant solutions are also required. To increase efficiency during various network issues, [23–25] presents Maximum Likelihood Event Localization (MLEM), fault tolerance employing repair points in clusters, and Failure Clustering Topology (FTCT). These models are further augmented by the use of component failures fault-tolerant models (NLFFT) with enhanced quadratic minimal level spanning tree [26], static fallback and dynamic timing tracking (SBDTM) [27], Kuatz Fault Tolerance [28], and Responsive Fault Tolerance (AFT) [29] models, which aid in network fault reduction by expansion of different networking parameters.

Explore model improvements for increased fault tolerance performance in real-time contexts, such as the use of Dual Cluster Heads Cooperation (CoDuch), the Delay Aware Regional Fault-Tolerant Routing Model, and distributed node categorization [30–32]. Because of the use of exceedingly complicated computational models, these models perform poorly. This performance can be enhanced by using discrete particle swarm optimization (DPSO), which is frequently based on real-time fault-tolerant task allocation algorithms (FTAOA) [33], collaborative sensor fault detection (CSFD), and Energy-Efficient and Cooperative Fault-Tolerant Communications, which aid in estimated network-level faults through pattern analysis. Nevertheless, none of these models combine fault tolerance with energy harvesting, limiting their practical use. The next part suggests creating a unique Hybrid Bio-inspired Model for Fault-Tolerant Energy Harvesting Networks with Fuzzy Rule Checks to improve its utility [34]. The proposed model was tested in a variety of network scenarios, and its performance was compared to that of other tested models for real-time analysis.

4.4 DESIGN OF A HYBRID BIO-INSPIRED MODEL FOR FAULT-TOLERANT ENERGY

According to the literature study, researchers provided a wide range of energy harvesting models with fault tolerance. To enhance their security and service quality metrics, each of these models has developed bespoke machine learning solutions. Despite this, these models require ongoing training, which raises their computation cost, storage cost, and end-to-end processing time, as well as reducing QoS availability for critical use cases.

To address these shortcomings, this chapter suggests an innovative lightweight model that uses evolutionary algorithms (GA) for range and energy-aware node clustering, a PSO for accident and fuel cells, and a hybrid ACO model with TLBO to aid in value-based fault forecasting and mitigation operations. The entire flow of the suggested model, which incorporates numerous bio-inspired models and accompanying process flows, is depicted in Figure 4.4.

The graph depicts how GA performs initial clustering, which assists in latency and energy-aware node grouping. These nodes are then routed through a fault tolerance model based on PSOs, which assists in finding energy-harvesting node sets. These node sets are submitted to the ACO and TLBO models for value-based fault testing, which aids in the improvement of QoS in a variety of communication contexts. The suggested model's design is separated into several sub-modules, which are detailed in distinct sections of this page. Readers will be able to implement these models in part(s) or whole based on their network requirements.

4.4.1 DESIGN OF THE GA LAYER FOR INITIAL CLUSTERING

On network router nodes, several databases, including network, node, and model databases, are preserved. Each of these databases contains an entity-specific collection of information that assists with parameter identification. A network database, for example, stores data on current bandwidth, data rate, received signal strength indicator (RSSI), capacity, temporal node performance, connection quality, and other network deployment metrics. The node database, on the other hand, maintains node-level data such as approximate locations, energy levels, energy models, node capacity, and so on. Decision data such as routing rules, energy threshold levels, and so on are stored in the model database. This information is sent into GA Model, which assists in the estimation of initial clusters. The following approach is used to achieve this clustering:

- **Initialize GA Parameters:**
    - Number of iterations (N_i);
    - Number of solutions (N_S);
    - Learning rate (L_r);
    - Maximum number of clusters ($C(Max)$).
- **Initially Mark Solution as to be Modified;**
- **For Each Iteration b/w 1 to N_i:**
    - For each solution b/w 1 to N_S.

- ➢ If the solution is marked as "not to be modified," then go to the next solution.
- ➢ Else, Generate a solution via the following process:

- Initialize number of clusters (N_c) stochastically via Equation 4.3:

$$Nc = STO_{CH}(2,\text{C (Max)}) \tag{4.3}$$

Where, STO_{CH} is stochastic Monrovian process for generating stochastic numbers for different sets.

- Based on this, divide nodes stochastically into Nc clusters via k Means, via evaluation of distance, energy, and node capacity-based comparison metric, which is evaluated via Equation 4.4,

$$d_k = \frac{E(Node) \times C(Node)}{(x(Node) - x(centroid))^2 + (y(Node - y(centroid))^2} \tag{4.4}$$

Where, d_k = distance metric for k Means, while $(Node)$ and $(Node)$ represents residual energy and capacity of node, and (x, y) represents current location of nodes respectively.

4.4.2 DESIGNING THE PSO LAYER FOR FAULT-TOLERANT ENERGY HARVESTING

Once nodes with similar energy, higher capacity, and a smaller distance from the centroid are aggregated, it is easier to identify clusters that may have defects during future communications. To accomplish this goal, a PSO-based model is utilized, which aids in the separation of node clusters depending on their temporal performance. The following is how this model (Figure 4.4) works:

- **Initialize PSO Parameters:**
  - ▪ Number of particles (Np);
  - ▪ Number of iterations (Ni);
  - ▪ Cognitive learning rate (Lc);
  - ▪ Social learning rate (Ls).
- **Determine Energy Model for Each Node, and Evaluate the Following Parameters:**
  - ▪ Energy needed for transmission ($E(T)$);

- Energy needed for reception ($E(R)$);
- Energy needed during idle mode ($E(I)$);
- Energy needed for waking up from sleep mode (E(W));
- Time needed for waking up from sleep mode (T(W)).

- **Based on This Model, Generate Initial Particles via the Following Process:**
  - For each cluster, evaluate cluster fitness via Equation 4.5 as follows;

$$C_f = \Sigma_{i=1}^{(N(n))})\frac{d_{k_i} - d_k(head}{N(n)} \times \left[\frac{T(W) \times E(Max)}{(E(T)+E(R)+E(I)+E(W)) \times Max(T(W))}\right] \quad (4.5)$$

  - For each particle in 1 to N_p, perform the following tasks;
  - Stochastically shift nodes with higher d_k values to clusters1 with lower C_f Evaluate particle velocities based on these shifts Via Equation 4.6,

$$PV(i) = \Sigma_{J=1}^{N_c} C_f(j) \times \frac{L_r}{N_c} \quad (4.6)$$

  - Mark the particle with the highest PV as $GBest$ via Equation 4.7,

$$GBest = Max[\cup_{i=1}^{N_p} PVi] \quad (4.7)$$

- **For Each Iteration in 1 to N_i, Perform Following Tasks:**
  - Update particle fitness via Equation 4.8:

$$(New) = \times (Old) + Lc \times [PV(Old) - PBest] + Ls[PV(Old) - \text{GBest}] \quad (4.8)$$

Where, r represents a stochastic number b/w range r ∈ (0.1, 1)

  - Based on this value of PV, shift nodes b/w clusters to obtain required particle velocities.
  - Update $PBest$, if $PBest > PV(New)$.
  - Similarly, update GBest via equation eight to obtain new particle sets.

- At end of final iteration, identify particle with highest velocity, and use its configuration for energy harvesting process.

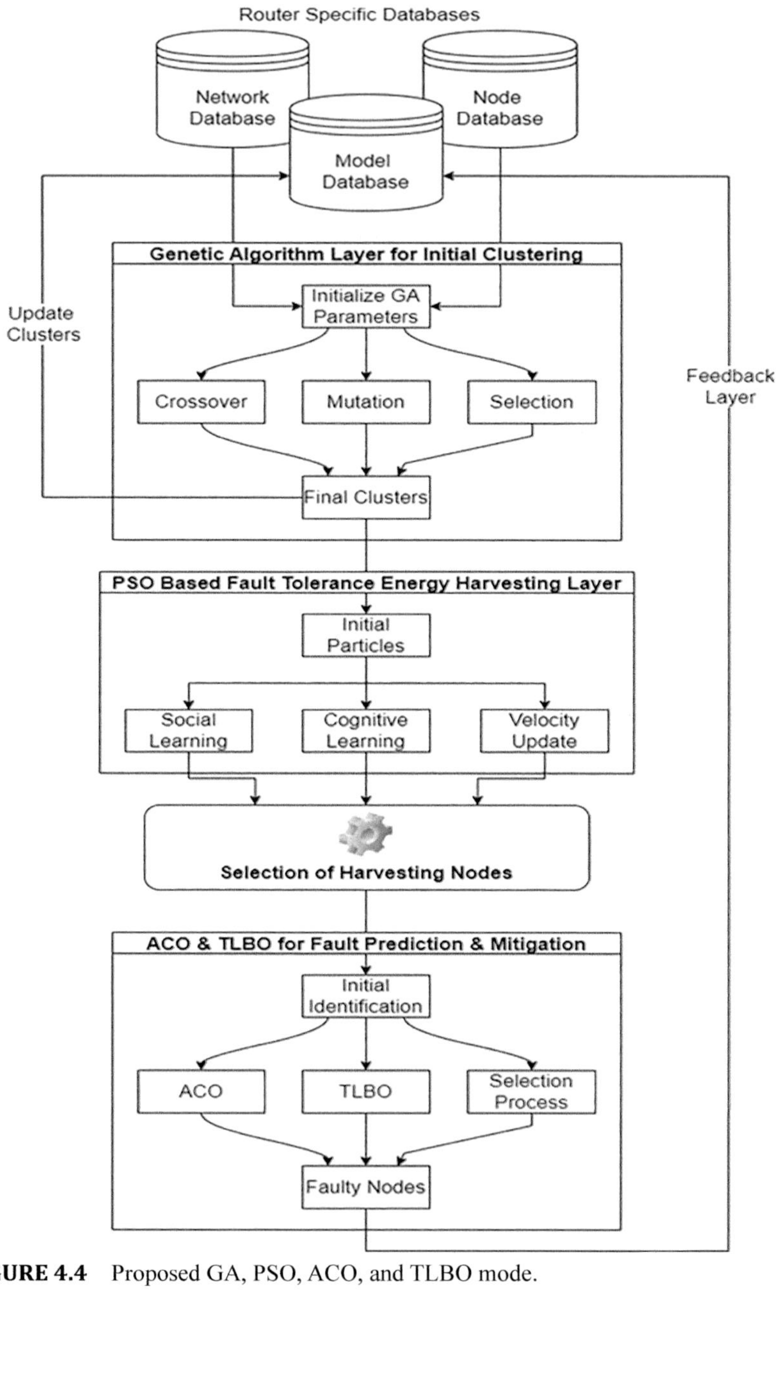

FIGURE 4.4 Proposed GA, PSO, ACO, and TLBO mode.

4.4.3 DESIGN OF THE HYBRID ACO AND TLBO MODEL

When non-faulty nodes are identified, an ACO and TLBO Model is implemented, which aids in value-based inspection of nodes for fault evaluation and mitigation throughout various communications. First, an ACO Model is utilized for each communication to discover the shortest path with the maximum residual energy using the following process:

- Initialize the following ACO parameters:
  - Number of ants (N_a).
  - Number of iterations (N_i).
  - Number of scout ants (N_s).
  - Pheromone regulation factor (α).
  - Visibility regulation factor (β).
  - Pheromone evaporation factor (γ).

- For each iteration in 1 to N_I
  - Pick Ns ants at random and evaluate node selection probability for each ant using Equation 4.9, as shown below:

$$p_{ij}^{l} = \frac{(\tau_{ij})^{\alpha} \times (d_{k_{ij}})^{\beta}}{\sum_{h=1}^{N(n)} (\tau_{ih})\alpha \times (d_{k_{ij}})^{\beta}} \tag{4.9}$$

Choose the indicated path for fault-tolerant routing with excellent QoS performance. A TLBO model is used to evaluate nodes in the chosen path, which aids in the discovery of fuzzy value changes during communications. The following is how this model works:

- Initialize TLBO Parameters:
  - Number of iterations (N_i);
  - Number of teachers ($N(T)$);
  - Number of learners (N_l).

- Teacher nodes are selected as nodes with higher energy levels in path, while other nodes are marked as learner nodes.
- For each iteration in 1 to N_i, perform the following tasks:
  - Select a stochastic Teacher Node, evaluate its sensor value, and convert this value into fuzzy range of Low (L=1), Medium (M=2), and High (H=3);

    - Determine the difference b/w the fuzzy values of any of the stochastic student nodes based on this fuzzy value;
    - If the difference b/w the fuzzy values is more than one, the student node is discarded and its defective count values are increased (FCV).
- At the end of all iterations, identify the fault count value threshold via Equation 10 as follows:

$$FCV_{th} = \sum_{i=1}^{N_l} FCV_i \times \frac{L_r}{N_l} \tag{4.10}$$

- If any node has $FCV > FCV_{th}$, then mark the node as "faulty" and shift it to "fault node cluster."

4.5 CONCLUSION

The suggested model optimizes various network parameters by combining several bio-inspired models. GA is used in this case for initial clustering, which assists in the identification of energy harvesting nodes. The PSO model supports this by analyzing malfunctioning nodes and using them as backups for the ongoing energy harvesting operation. These models are further strengthened by the use of an ACO-based TLBO model with fuzzy rule-based value checks, as well as delay-aware, energy-aware, and capacity-aware nodes for routes in dense wireless network environments.

KEYWORDS

- **ant colony optimization**
- **energy harvesting**
- **genetic algorithm**
- **particle swarm optimization**
- **quality of service**
- **teacher learner based optimization**
- **wireless sensor network**

REFERENCES

1. Mhatre, K. P., & Khot, U. P. (2020). Energy efficient opportunistic routing with sleep scheduling in wireless sensor networks. *Wireless Personal Communications, 112*(2), 1243–1263.
2. Deb, M., & Roy, S. (2021). Enhanced-Pro: A new enhanced solar energy harvested prediction model for wireless sensor networks. *Wireless Personal Communications, 117*, 1103–1121. https://doi.org/10.1007/s11277-020-07913-y.
3. Banerjee, P. S. (2020). Sleep: Thermal entropy aware intelligent sleep scheduling algorithm for wireless sensor network. *Microsystem Technologies: Sensors, Actuators, Systems Integration, 26*(7), 2305–2323.
4. Vu, H. S., Nguyen, N., Ha-Van, N., Seo, C., & Thuy Le, M. (2020). Multiband ambient RF energy harvesting for autonomous IoT devices. *IEEE Microwave and Wireless Components Letters, 30*(12), 1189–1192. https://doi.org/10.1109/LMWC.2020.3029869.
5. Luo, Y., & Pu, L. (2021). Practical issues of RF energy harvest and data transmission in renewable radio energy powered IoT. *IEEE Transactions on Sustainable Computing, 6*(4), 667–678. https://doi.org/10.1109/TSUSC.2020.3000085.
6. Wang, Y., Yang, K., Wan, W., Zhang, Y., & Liu, Q. (2021). Energy-efficient data and energy integrated management strategy for IoT devices based on RF energy harvesting. *IEEE Internet of Things Journal, 8*(17), 13640–13651.
7. Ren, Q., & Yao, G. (2021). Enhancing harvested energy utilization for energy harvesting wireless sensor networks by an improved uneven clustering protocol. *IEEE Access, 9*, 119279–119288.
8. Moon, K., Kim, K. M., Kim, Y., & Lee, T. J. (2021). Device-selective energy request in RF energy-harvesting networks. *IEEE Communications Letters, 25*(5), 1716–1719.
9. Ghosh, D., Hanawal, M. K., & Zlatanov, N. (2021). Learning to optimize energy efficiency in energy harvesting wireless sensor networks. *IEEE Wireless Communications Letters, 10*(6), 1153–1157.
10. Chew, Z. J., Ruan, T., & Zhu, M. (2020). Energy savvy network joining strategies for energy harvesting powered TSCH nodes. *IEEE Transactions on Industrial Informatics, 17*(2), 1505–1514.
11. Sun, Y., Song, C., Yu, S., Liu, Y., Pan, H., & Zeng, P. (2021). Energy-efficient task offloading based on differential evolution in edge computing system with energy harvesting. *IEEE Access, 9*, 16383–16391.
12. Khan, A., Khan, M., Ahmed, S., Iqbal, N., Abd Rahman, M. A., Karim, M. K. A., & Yaakob, Y. (2020). EH-IRSP: Energy harvesting based intelligent relay selection protocol. *IEEE Access, 9*, 64189–64199.
13. Huang, J., Yu, B., Xing, C. C., Cerny, T., & Ning, Z. (2020). Online energy scheduling policies in energy harvesting enabled D2D communications. *IEEE Transactions on Industrial Informatics, 17*(8), 5678–5687.
14. Jaitawat, A., & Singh, A. K. (2020). Online transmission policy for energy harvesting sensor node with energy loss. *IEEE Communications Letters, 25*(2), 551–554.
15. Cheng, Y., Xia, W., Zhao, H., Yang, L., & Zhu, H. (2021). Joint caching and user association for energy harvesting aided Internet of Things with full-duplex backhauls. *Journal of Communications and Information Networks, 6*(4), 420–428.

16. Cui, E., Yang, D., Zhang, H., & Gidlund, M. (2020). Improving power stability of energy harvesting devices with edge computing-assisted time fair energy allocation. *IEEE Transactions on Green Communications and Networking, 5*(1), 540–551.
17. Hoseinghorban, A., Bahrami, M. R., Ejlali, A., & Abam, M. A. (2020). CHANCE: Capacitor charging management scheme in energy harvesting systems. *IEEE Transactions on Computer-Aided Design of Integrated Circuits and Systems, 40*(3), 419–429.
18. Lee, J. S., & Jiang, H. T. (2020). An extended hierarchical clustering approach to energy-harvesting mobile wireless sensor networks. *IEEE Internet of Things Journal, 8*(9), 7105–7114.
19. Abdul-Qawy, A. S. H., Nasser, A. B., Guroob, A. H., Saad, A. M. H., Alduais, N. A. M., & Khatri, N. (2021). TEMSEP: Threshold-oriented and energy-harvesting enabled multilevel SEP protocol for improving energy-efficiency of heterogeneous WSNs. *IEEE Access, 9*, 154975–155002.
20. Stai, E., & Karyotis, V. (2020). Optimal resource allocation in multihop wireless networks relying on energy harvesting. *IEEE Communications Letters, 25*(1), 224–228.
21. Ali, K., & Rogers, D. J. (2020). An orientation-independent multi-input energy harvesting wireless sensor node. *IEEE Transactions on Industrial Electronics, 68*(2), 1665–1674.
22. Wang, T., Wu, S., Wang, Z., Jiang, Y., Ma, T., & Yang, Z. (2020). A multi-featured actor-critic relay selection scheme for large-scale energy harvesting WSNs. *IEEE Wireless Communications Letters, 10*(1), 180–184.
23. Dash, D. (2023). Plane sweep algorithms for data collection for energy harvesting wireless sensor networks using mobile sink. *Journal of Ambient Intelligence and Humanized Computing, 14*(5), 6307–6320.
24. Al-Qamaji, A., & Atakan, B. (2022). Event distortion-based clustering algorithm for energy harvesting wireless sensor networks. *Wireless Personal Communications, 123(4), 1–21.*
25. Panayiotou, C. G., & Michaelides, M. P. (2009). Fault tolerant maximum likelihood event localization in sensor networks using binary data. *Proceedings of the American Control Conference, 58*, 3151–3156.
26. Bhuiyan, M. Z. A., Wang, G., Cao, J., & Wu, J. (2013). Deploying wireless sensor networks with fault-tolerance for structural health monitoring. *IEEE Transactions on Computers, 64*(2), 382–395.
27. Hu, S., & Li, G. (2018). Fault-tolerant clustering topology evolution mechanism of wireless sensor networks. *IEEE Access, 6*, 28085–28096.
28. Menaria, V. K., Jain, S. C., Raju, N., Kumari, R., Nayyar, A., & Hosain, E. (2020). NLFFT: A novel fault tolerance model using artificial intelligence to improve performance in wireless sensor networks. *IEEE Access, 8*, 149231–149254.
29. Tong, Y., Tian, L., Lin, L., & Wang, Z. (2020). Fault tolerance mechanism combining static backup and dynamic timing monitoring for cluster heads. *IEEE Access, 8*, 43277–43288.
30. Shen, H., & Li, Z. (2015). A Kautz-based wireless sensor and actuator network for real-time, fault-tolerant and energy-efficient transmission. *IEEE Transactions on Mobile Computing, 15*(1), 1–16.

31. Chen, R., Speer, A. P., & Eltoweissy, M. (2010). Adaptive fault-tolerant QoS control algorithms for maximizing system lifetime of query-based wireless sensor networks. *IEEE Transactions on Dependable and Secure Computing, 8*(2), 161–176.
32. Huang, G., Zhang, Y., He, J., & Cao, J. (2011). Fault tolerance in data gathering wireless sensor networks. *The Computer Journal, 54*(6), 976–987.
33. Ouyang, Y., Wang, Q., Ru, M., Liang, H., & Li, J. (2020). A novel low-latency regional fault-aware fault-tolerant routing algorithm for wireless NoC. *IEEE Access, 8*, 22650–22663.
34. Wang, T. Y., Han, Y. S., & Varshney, P. K. (2005). Fault-tolerant distributed classification based on non-binary codes in wireless sensor networks. *IEEE Communications Letters, 9*(9), 808–810.
35. Guo, W., Li, J., Chen, G., Niu, Y., & Chen, C. (2014). A PSO-optimized real-time fault-tolerant task allocation algorithm in wireless sensor networks. *IEEE Transactions on Parallel and Distributed Systems, 26*(12), 3236–3249.

CHAPTER 5

Design of Notch and Slot Loaded Rectangular Microstrip Patch Antenna for IoT/ISM/WLAN Applications in S-Band

RAMESH KUMAR VERMA,[1,*] AKHILESH KUMAR,[2] PRABINA PATTANAYAK,[2] AJAY KUMAR,[3] VIKRAM BALI,[1] and SAURABH DWIVEDI[1]

[1]*Department of Computer Science and Engineering, IMS Engineering College, Ghaziabad, Uttar Pradesh, India*

[2]*Department of Electronics and Communication Engineering, National Institute of Technology, Silchar, Assam, India*

[3]*Department of Electronics and Communication Engineering, IMS Engineering College, Ghaziabad, Uttar Pradesh, India*

*Corresponding author

ABSTRACT

In this chapter, a slot-loaded compact rectangular microstrip antenna has been designed by cutting a rectangular slot and various notches of different sizes. The impedance bandwidth of the suggested microstrip antenna is 320 MHz, and the antenna is resonating at 2.339 GHz. The suggested antenna structure has a frequency band from 2.16 GHz–2.48 GHz. This frequency band is appropriate for IoT, ISM, WLAN, and other wireless communication applications in the S-band. The return loss of the suggested antenna is –31.10 dB at the resonant frequency of 2.339 GHz. The gain and antenna efficiency of the suggested antenna are 3.54 dB and 99.92%, respectively,

Wireless Technology: Advances and Innovations for Smart IoT Applications. Rajesh Kumar Dwivedi, Vinod Kumar Singh, and Raghav Dwivedi (Eds.)

DOI: 10.1201/9781998511495-5

at the resonant frequency of 2.339 GHz. The suggested microstrip antenna is excited by a 50Ω microstrip line feed. The suggested antenna has been simulated using the IE3D simulation tool.

5.1 INTRODUCTION

The advancement in wireless communication devices has expanded the interest in compact and low-profile antennas with improved bandwidth and high gain. The microstrip antennas have a lot of advantages such as low profile, light weight, and compactness, but the main disadvantages of microstrip antennas are their low efficiency, narrow frequency bandwidth, spurious feed radiation, and lower gain. There are numerous substrates whose dielectric constants lie in the range of 2.2–12 that can be utilized for the design of microstrip antennas. The substrate with a lower dielectric constant gives better efficiency and a larger impedance bandwidth [1].

The bandwidth of a microstrip patch antenna can be raised by utilizing various strategies, for example, taking different shapes of patches, loading the patch with various types of slots and notches [2]. A hexagonal shape slotted rectenna is presented for Wi-Fi applications using a hexagonally shaped microstrip radiator operating at 2.45 GHz [3], while a stub and notches loaded equivalent circuit-based antenna is designed for wireless applications [4]. A slotted and compact patch antenna design operating at a frequency of 3.30 GHz is proposed for S-band applications [5]. A compact design having a semi-circular radiating patch is presented for triple band using defected ground for C/X/Ku/K-band applications [6], while a six-port MIMO antenna resonating at 4.8 GHz is designed for C-band operations [7]. However, an array antenna [8] for 5G sub-6 GHz and a fractal antenna [9] using a ground defected structure are also presented. Another two antennas having T-shape slots [10] and plus shape slots [11] have been designed and optimized using PSO. Furthermore, an inverted pentagonal shape CDRA for dual band [12] and two-layered rectangular patch antennas [13] are also proposed. However, a wearable UWB antenna of parachute shape using a jeans textile substrate for biomedical applications [14] and a dual band flexible textile antenna using jeans fabric for WiMAX and WLAN applications [15] are designed. Another UWB antenna based on circuit theory with band notch characteristics is also presented [16]. A planar antenna with dual band characteristics is presented for GSM and

WiMAX using FR-4 substrate [17], while a DRA-based MIMO antenna with four ports is designed for sub-6 GHz 5G applications [18]. Furthermore, a circularly polarized ring CDRA using an inverted L-shaped slotted feed line with a tilted rectangular shape aperture [19] and a concentric annular ring with symmetrical notches and a cross shape slot [20] are presented. However, a gap-coupled antenna for wideband [21] and an antenna array for X-band applications [22] are also designed.

In the current research work, a microstrip patch antenna has been presented by cutting slots and notches in the radiating patch, which is fed by a 50Ω microstrip line feed. The impedance bandwidth of the suggested antenna has been achieved at 320 MHz, which lies between 2.16 GHz and 2.48 GHz. The presented antenna is resonating at 2.339 GHz with a –31.10 dB return loss. The resonating band of the antenna is appropriate for IoT, ISM, WLAN, and other communication applications [23].

5.2 ANTENNA DESIGN

For and antenna design, the length and width of rectangular patch are evaluated by using Equations 5.1–5.4 [24].

$$W = \frac{c}{2f_r}\sqrt{\frac{2}{\varepsilon_r + 1}} \tag{5.1}$$

where $c = 3\times10^8$ m/s (speed of light) in vacuum, $\varepsilon_r = 4.4$ (dielectric constant) of substrate, f_r = 2.50 GHz (design frequency), W is the width of patch, and ε_{reff} is effective dielectric constant given as:

$$\varepsilon_{reff} = \frac{\varepsilon_r + 1}{2} + \frac{\varepsilon_r - 1}{2}\left(1 + \frac{12h}{W}\right)^{-\frac{1}{2}} \tag{5.2}$$

Extension length (ΔL) of patch is evaluated by:

$$\frac{\Delta L}{\mathrm{h}} = 0.412\frac{(\varepsilon reff + 0.3)(\frac{W}{h} + 0.264)}{(\varepsilon reff + 0.258)(\frac{W}{h} + 0.8)} \tag{5.3}$$

Equation 5.3 can be used to evaluate the value of absolute length of the patch as:

$$L = \frac{c}{2fr\sqrt{\varepsilon_{reff}}} - 2\Delta L \tag{5.4}$$

Patch length (L_p) and width (W_p) of suggested antenna has been evaluated by Equations 5.1–5.4. The evaluated patch length and width are 28.24 mm and 36.50 mm respectively. Ground length and width can be evaluated by Equations 5.5 and 5.6 as given below [24].

$$Lg = 6h + Lp \tag{5.5}$$

$$Wg = 6h + Wp \tag{5.6}$$

5.3 DESIGN SPECIFICATIONS OF ANTENNA

The suggested antenna has been planned by utilizing glass epoxy substrate material with a dielectric constant of 4.4 at a design frequency of 2.50 GHz. The height of the dielectric substrate is 1.6 mm, with a loss tangent of 0.02 [25]. The patch length and width of the suggested antenna have been evaluated using Equations 5.1–5.4. The evaluated patch length (horizontal) and width (vertical) are 28.24 mm and 36.50 mm, respectively. The rectangular ground length and width have been evaluated using Equations 5.5 and 5.6. The suggested antenna has been designed for a finite ground. For the evaluation of the ground length and width of the suggested antenna, *6h* has been added to both the length and width of the patch. The evaluated ground length and width are 37.84 mm and 46.10 mm, respectively. The design specifications of the proposed antenna are shown in Table 5.1. The suggested antenna is excited through a 50Ω microstrip line feed. Simulation and analysis work have been performed using the IE3D simulation tool [26].

TABLE 5.1 Design Specifications of Antenna

Parameters	Value	Parameters	Value
Design frequency f_r	2.50 GHz	Patch length L_p	28.24 mm
Dielectric constant ε_r	4.4	Patch width W_p	36.50 mm
Substrate height h	1.6 mm	Ground plane length L_g	37.84 mm
Loss tangent δ	0.02	Ground plane width W_g	46.10 mm

5.4 GEOMETRY AND DESIGN PROCEDURE OF ANTENNA

In the process of designing the proposed antenna using IE3D software, careful attention was given to the placement of the ground plane, situated at coordinates (0, 0), positioned at the lower-left corner of the ground. The design of the microstrip patch antenna involved the incorporation of a slot and notches of various sizes on the antenna's patch. The intricacies of the suggested antenna's geometry have been visually represented in Figure 5.1, providing a comprehensive illustration of its structural composition. Furthermore, the specific dimensions crucial to the antenna's construction have been meticulously outlined and detailed in Table 5.2. Notably, the microstrip line feed, adhering to a 50 Ω impedance, was strategically connected to the lower-left edge of the patch via a strip characterized by a length of 5 mm and a width of 4.8 mm. This deliberate configuration ensures optimal functionality and performance of the antenna, aligning with the intended specifications of the design.

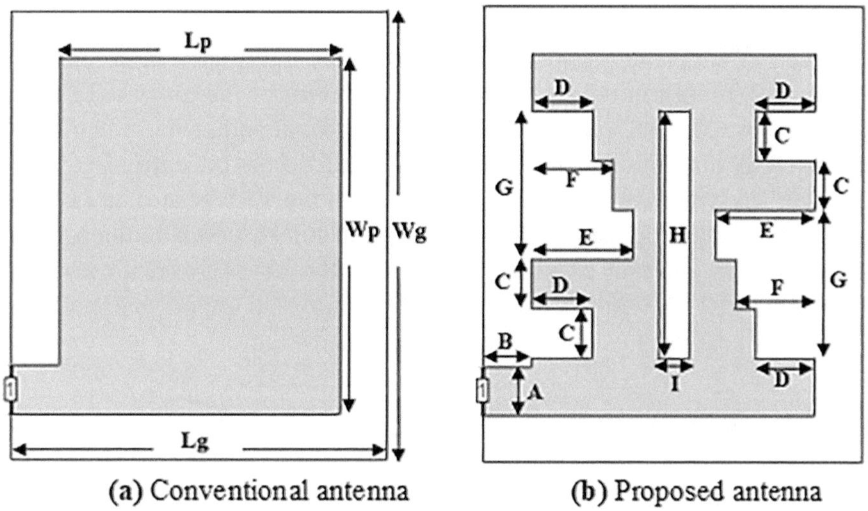

FIGURE 5.1 Geometry of proposed antenna.

TABLE 5.2 Slot and Notch Parameters of Antenna

Parameters	A	B	C	D	E	F	G	H	I
Value (mm)	5	4.8	5	6	10	8	15	25	3

5.5 RESULTS AND DISCUSSION

In the current research endeavor, the enhancement of the bandwidth of a rectangular slot-loaded microstrip patch antenna was pursued through the incorporation of notches of varying sizes and a slot characterized by a length of 25 mm and a width of 3 mm, as outlined in Table 4.2. The comprehensive analysis of the suggested microstrip patch antenna was conducted through simulation using the IE3D software, spanning a frequency range from 1 GHz–3 GHz with 1000 frequency points. The ensuing performance evaluation encompassed critical parameters such as return loss, Voltage Standing Wave Ratio (VSWR), Smith chart and Z-parameter, gain, efficiency, and radiation pattern, all meticulously documented in Figures 5.2–5.7. The impedance bandwidth of the conventional antenna was determined to be 280 MHz, spanning from 2.33 GHz–2.61 GHz, resonating at 2.48 GHz with a return loss of –15.58 dB. In contrast, the proposed antenna exhibited an expanded impedance bandwidth of 320 MHz, ranging from 2.16 GHz–2.48 GHz, resonating at 2.339 GHz with an impressive return loss of –31.10 dB. The comparative return loss versus frequency graph for both antennas is visually represented in Figure 5.2, highlighting the superior performance of the proposed design. Furthermore, the VSWR for both the conventional and proposed antennas consistently maintained values between 1 and 2 across the entire frequency band, as illustrated in Figure 5.3. Specifically, the VSWR measured 1.339 at the resonant frequency of 2.48 GHz for the conventional antenna, while the proposed antenna exhibited a lower value of 1.057 at its resonant frequency of 2.339 GHz, underscoring the improved impedance matching and overall efficacy of the suggested antenna design.

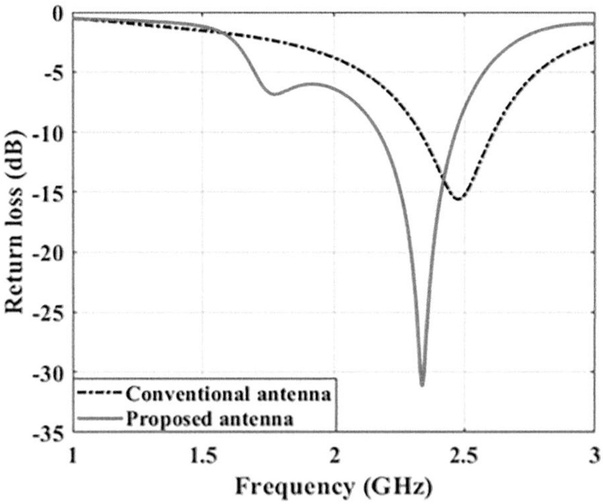

FIGURE 5.2 Return loss Vs frequency graph of conventional and proposed antenna.

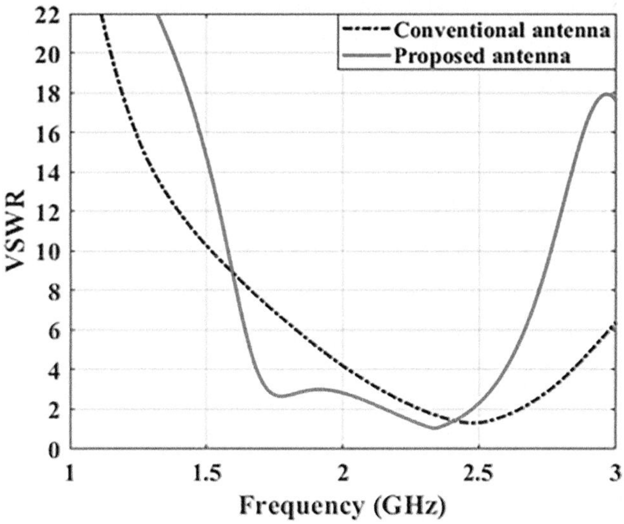

FIGURE 5.3 VSWR graph of conventional and proposed antenna.

The intricate analysis of the suggested antenna extends to the Smith chart and Z-parameter, as depicted in Figure 5.4. The Z-parameter for the suggested antenna has been meticulously determined, yielding a value of Z = 47.29–j0.25 Ω at the resonant frequency of 2.339 GHz. This information is crucial in understanding the impedance characteristics and performance of the antenna at its peak resonance. Moving forward, the gain and antenna efficiency of the suggested antenna are graphically presented in Figures 5.5 and 5.6, respectively. The peak gain, a pivotal indicator of the antenna's directional radiation capabilities, is recorded at 3.54 dB, showcasing its effectiveness in transmitting signals in the specified frequency range. Simultaneously, the antenna efficiency is exceptionally high, reaching 99.92% at the resonant frequency of 2.339 GHz, underscoring the efficacy of the antenna in converting input power into radiated energy. Furthermore, the 2D radiation pattern of the suggested antenna is vividly illustrated in Figure 5.7, providing insights into its directional characteristics. The 3 dB beam width, a measure of the angular spread of the antenna's radiation pattern, is delineated by coordinates (50.2, 136.63) degrees, offering valuable information on the spatial coverage and focus of the antenna's radiated energy. This comprehensive evaluation not only validates the proposed antenna's superior performance in terms of impedance matching, gain, and efficiency but also provides essential insights into its directional radiation behavior, further substantiating its potential utility in practical applications.

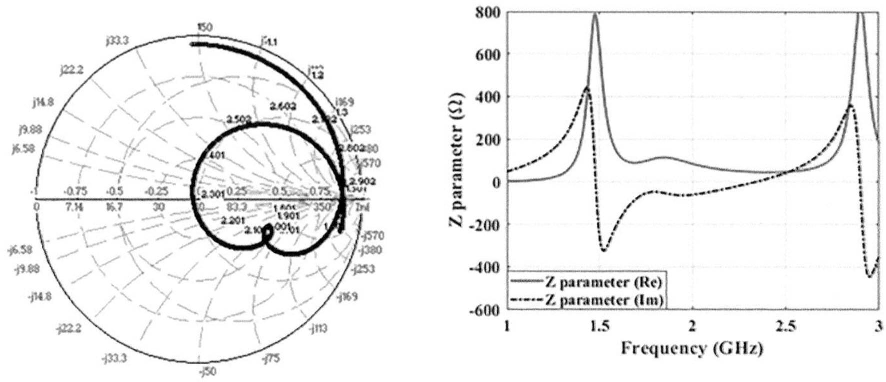

a b

FIGURE 5.4 Smith chart and Z-parameter of suggested antenna.

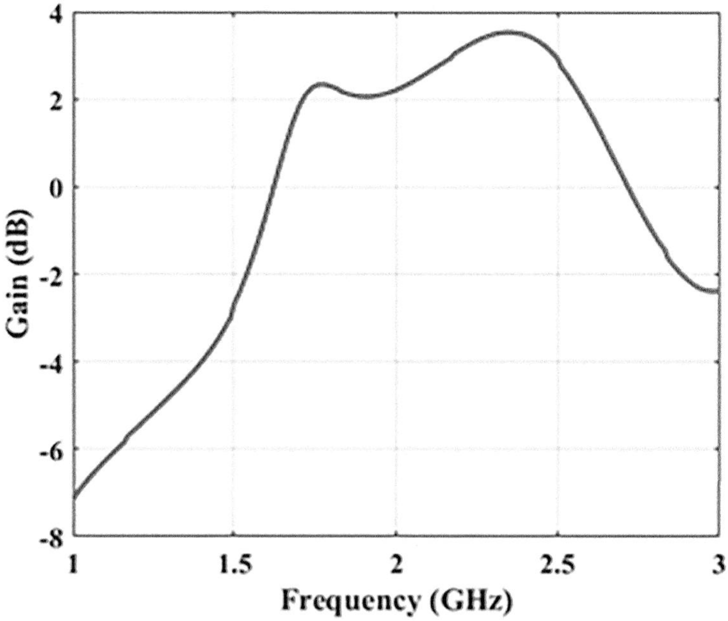

FIGURE 5.5 Gain Vs frequency graph of suggested antenna.

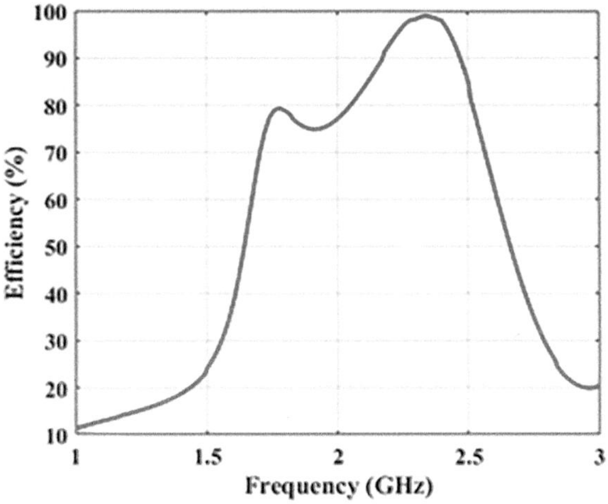

FIGURE 5.6 Antenna efficiency Vs frequency graph of suggested antenna.

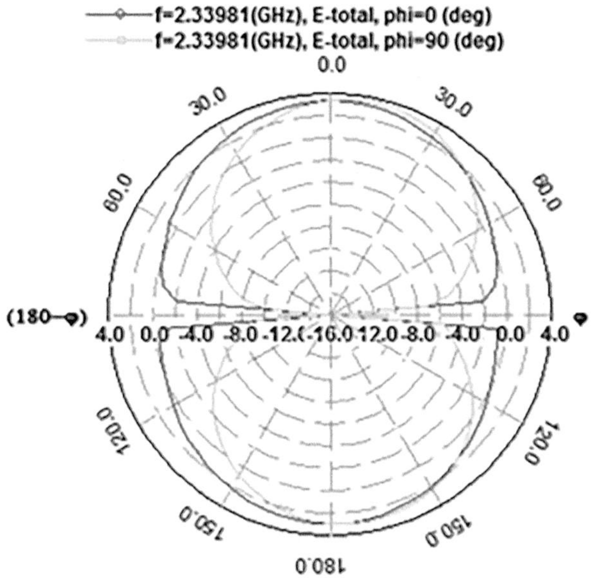

FIGURE 5.7 Radiation pattern (2D) of suggested antenna at 2.339 GHz.

5.6 CONCLUSION

The design and analysis of the suggested antenna have been studied. The impedance bandwidth of a conventional microstrip antenna is just a couple of percent. Thus, it becomes very essential to develop a new design and technique to improve the bandwidth of the microstrip patch antenna. The suggested antenna improved the impedance bandwidth to 320 MHz and is resonating at 2.339 GHz. The suggested antenna has been designed on a glass epoxy (FR4) substrate material to achieve a peak antenna efficiency of 99.92% and a gain of 3.54 dB at the resonant frequency of 2.339 GHz.

KEYWORDS

- **bandwidth**
- **line feed**
- **microstrip**
- **patch antenna**
- **radiation**
- **S-band**

REFERENCES

1. Balanis, C. A. (2016). *Antenna Theory: Analysis and Design*. John Wiley & Sons.
2. Mishra, B., Verma, R. K., Yashwanth, N., & Singh, R. K. (2022). A review on microstrip patch antenna parameters of different geometry and bandwidth enhancement techniques. *International Journal of Microwave and Wireless Technologies, 14*(5), 652–673.
3. Dhar, A., Pattanayak, P., Kumar, A., Gurjar, D. S., & Kumar, B. (2022). Design of a hexagonal slot rectenna for RF energy harvesting in Wi-Fi/WLAN applications. *International Journal of RF and Microwave Computer-Aided Engineering, 32*(12), e23512.
4. Verma, R. K., Srivastava, D. K., & Mishra, B. (2022). Circuit theory model-based analysis of triple-band stub and notches loaded epoxy substrate patch antenna for wireless applications. *International Journal of Communication Systems, 35*(4), e5052.
5. Tripathi, D., Srivastava, D. K., & Verma, R. K. (2021). Bandwidth enhancement of slotted rectangular wideband microstrip antenna for the application of WLAN/WiMAX. *Wireless Personal Communications, 119*, 1193–1207.
6. Kumar, A., Pattanayak, P., & Dhar, A. (2023). Compact triple band microstrip patch antenna for satellite and C/X/K/Ku bands applications. *Wireless Personal Communications, 129*(1), 57–70.

7. Kumar, A., Mohanty, J., Pattanayak, P., Sabat, D., & Prasad, G. (2023). Six-port mid-bands Low-SAR MIMO antenna for WLAN, 5G mobile terminals, and C-band applications. *AEU-International Journal of Electronics and Communications, 166*, 154665.
8. Raj, A., & Mandal, D. (2024). Design and implementation of hybrid fed array antennae for Sub-6 GHz and 5G mm-wave communication and wireless applications. *International Journal of Microwave and Wireless Technologies*, 1–20.
9. Raj, A., & Mandal, D. (2024). Design and experimental analysis of fractal antennae with ground-defected structure for expected 6G and 5G mm-wave communication and wireless applications. *Transactions on Emerging Telecommunications Technologies, 35*(1), e4900.
10. Verma, R. K., & Srivastava, D. K. (2019). Design, optimization and comparative analysis of T-shape slot loaded microstrip patch antenna using PSO. *Photonic Network Communications, 38*, 343–355.
11. Verma, R. K., & Srivastava, D. K. (2021). Optimization and parametric analysis of slotted microstrip antenna using particle swarm optimization and curve fitting. *International Journal of Circuit Theory and Applications, 49*(7), 1868–1883.
12. Rai, C., Singh, A., Singh, S., Singh, A. K., & Verma, R. K. (2022). Dual-band and dual polarized inverted pentagonal shaped hybrid cylindrical dielectric resonator antenna for wireless applications. *Wireless Personal Communications*, *124,* 2121–2139.
13. Choudhary, S. D. (2023). Design of microstrip rectangular patch antenna using coplanar parasitic rod elements with two-layer substrate coupled integrated feeding line technique. *Wireless Personal Communications, 131*(4), 3073–3087.
14. Singh, V. K., Bangari, N., Ali, Z., Vyas, A., Verma, R. K., & Saxena, A. (2023). Parachute shape ultra-wideband wearable antenna for remote health care monitoring. *International Journal of Communication Systems, 36*(10), e5488.
15. Singh, V. K., Dhupkariya, S., & Bangari, N. (2017). Wearable ultra wide dual band flexible textile antenna for WiMax/WLAN application. *Wireless Personal Communications, 95*(2), 1075–1086.
16. Yadav, A., Singh, P., Verma, R. K., & Singh, V. K. (2023). Design and comparative analysis of circuit theory model-based slot-loaded printed rectangular monopole antenna for UWB applications with notch band. *International Journal of Communication Systems, 36*(3), e5390.
17. Patel, U., & Upadhyaya, T. K. (2020). Dual band planar antenna for GSM and WiMAX applications with inclusion of modified split ring resonator structure. *Progress In Electromagnetics Research Letters, 91*, 1–7.
18. Patel, U., & Upadhyaya, T. (2022). Four-port dual-band multiple-input multiple-output dielectric resonator antenna for Sub-6 GHz 5G communication applications. *Micromachines, 13*(11), 2022.
19. Rai, C., Singh, S., Singh, A. K., & Verma, R. K. (2022). A novel design of circularly polarized ring CDRA with wideband impedance bandwidth using slotted microstrip feed line for X-band applications. *International Journal of Microwave and Wireless Technologies, 14*(9), 1149–1158.
20. Choudhary, S. D., Patil, S., Verma, A., Alam, M. I., Kapse, V. M., & Kanaujia, B. K. (2022). Design of dual-polarized triple-band concentric annular-ring microstrip patch

antenna for GPS applications. *International Journal of Microwave and Wireless Technologies, 14*(10), 1338–1346.

21. Gupta, A., Srivastava, D. K., Saini, J. P., & Verma, R. K. (2020). Comparative analysis of microstrip-line-fed gap-coupled and direct-coupled microstrip patch antennas for wideband applications. *Journal of Computational Electronics, 19*, 457–468.
22. Saxena, A., & Singh, V. K. (2022). Design of compact array antenna and its effect on the human brain. *Wireless Personal Communications, 125*(1), 637–647.
23. Verma, R. K. (2023). Design, analysis and fabrication of compact and T-shape notches loaded dual-band rectangular microstrip patch antenna for GPS/WLAN/WiMAX applications. *Wireless Personal Communications, 132*(1), 505–522.
24. Verma, R. K. (2024). Bandwidth enhancement of compact and wideband square microstrip patch antenna for WLAN/WiMAX applications in C-band. *International Journal of Numerical Modelling: Electronic Networks, Devices and Fields, 37*(2), e3178.
25. Verma, R. K., Priya, B., Singh, M., Singh, P., Yadav, A., & Singh, V. K. (2023). Equivalent circuit model-based design and analysis of microstrip line-fed electrically small patch antenna for sub-6 GHz 5G applications. *International Journal of Communication Systems, 36*(17), e5595.
26. IE3D Electromagnetic Simulation and Optimization Package, Version 9.0.

CHAPTER 6

Design of Slotted Hammer-Type Stretchable Antenna for IoT Applications

ZAKIR ALI,[1*] SHASHIKANT VERMA,[2] GAURAV VERMA,[2] and SAIYED TAZEN ALI[1]

[1]*Department of Electronics and Telecommunication Engineering, IET, Bundelkhand University, Jhansi, Uttar Pradesh, India*

[2]*Department of Electronics and Communication Engineering, IET, Jhansi, Uttar Pradesh, India*

*Corresponding author

ABSTRACT

These days, consumers choose products that are very portable and lightweight rather than bulky ones. As a result, in order to satisfy the requirements of customers, a triple-band stretchable antenna has been developed for wireless applications and low-power IoT devices, and it has been implemented on a flexible substrate. The antenna described here is quite small. The planned antenna will have a resonant frequency of 11.43 GHz. The highest directivity of the antenna, as calculated by simulation, is 5.739 dBi. The antenna being exhibited has a triple bandwidth of 20.18%, 17.05%, and 40.09%. Because the planned antenna will have a modest profile, it will be particularly well-suited for use in applications related to the Internet of Things. It is well-suited for use in Internet of Things applications.

Wireless Technology: Advances and Innovations for Smart IoT Applications. Rajesh Kumar Dwivedi, Vinod Kumar Singh, and Raghav Dwivedi (Eds.)

DOI: 10.1201/9781998511495-6

6.1 INTRODUCTION

The vast majority of flexible antenna applications are well suited for use in healthcare and telemedicine; the reason for this lies in the antenna's ability to be both flexible and small. Because of their relatively high gain and bandwidth, small antennas play an important part in the area of satellite communication as well as mobile communication [6–11]. Applications of the Internet of Things are responsible for the most significant advancements in the area of current communication systems. It is essential that the devices used for effective and dependable communication be small in size, have little or no production costs, a broad bandwidth, and high energy efficiency. Today's Internet of Things (IoT) devices are very well suited for use with LTE, WLAN, and WiMAX applications. As the global population continues to expand at an exponential rate, there will be an accompanying increase in the need for small devices that also have flexible and lightweight antenna modules. Because embedded systems are so compact, it is possible to integrate many modules on relatively tiny devices like these in order to boost operational efficacy, dependability, and resilience across a wide range of use cases, including environmental monitoring, smart cities, smart healthcare, smart grids, military, and defense, and so on [1, 2]. It is projected that the antenna system, which is the front end of all hand-held communication devices, will cover all main frequency bands of IEEE 802.15.4 (2.5 GHz–2.69 GHz) and IEEE 802.11 (2.4 GHz–2.48 GHz) with adequate gain and radiation patterns for a variety of integrated services. This will be possible since the antenna system is the front end of all hand-held communication devices. In addition, it is anticipated that modern antenna design will be flexible enough to enable independent adjustment of impedance bandwidth for a variety of center frequencies [4]. Some well-known strategies have been used in order to fulfill the aforementioned criteria. These strategies include cutting holes into the radiating patch, using defective ground structures (DGS), engraving strips on antennas, and incorporating band-notched structures into design [4, 5]. In addition to these more traditional methods, the published research also explores the use of metamaterials and complementary split ring resonator (CSRR) approaches, with the goals of achieving better gain and reducing cross-frequency interference. In their experiment, Ahmed et al. employed a magneto-electric (ME) dipole antenna, which they showed had a broad impedance bandwidth, enhanced gain, and matched radiation patterns on both the E-plane and the H-plane [6]. Cross-magneto-electric structures are not the best candidates for mass manufacturing in the IoT due to their size and the sensitive design constraints they impose. There have been several CPW-fed antennas recorded in the literature. These antennas include multiband [3–7], CPW antennas with extra strips for WLAN [8–10], and asymmetric coplanar strip antennas [11–16]. However, the vast majority of these designs have huge antennas and do not cover all of the important bands of the electromagnetic spectrum. A dual-band antenna that measures 25 mm by 25 mm and has an average gain of 2.5 dB and performs only in the WLAN band. This antenna only operates in the WLAN band. In addition, the antenna gains and reflection coefficient parameters do not impressively fit the antenna dimensions in certain other designs [14–16]. These designs may be identified by the fact that the antenna gains are not stunning. Only on the WiMAX band does make use of a CPW-supplied multiband antenna with dimensions of 70.4 mm by 45 mm and an impedance bandwidth of 127 MHz. An extremely low impedance bandwidth is present in a 70 mm × 70 mm CPW-fed WLAN antenna. This antenna has a peak gain of 6.5 dB, and it works at 2.4 GHz. In the research that has been done, the concept of "rounded corners" has been brought up as a way to increase total gain and assure stable radiation patterns.

6.2 ANTENNA DESIGN PROCEDURE

The antenna shown in Figure 6.1 has a low profile and is relatively small in size, measuring 49.5 mm by 46.7 mm. After computing the length and breadth of the rectangular substrate, the acquired dimensions are utilized to provide a triple bandwidth. Although there are many other feeding strategies available, the line feed approach is the one used in this chapter to achieve a very wide bandwidth and gain. The CST software is used for modeling and simulation of the proposed antenna, which provides triple bandwidth. This may be achieved by utilizing the antenna's working range. Table 6.1 presents the design parameters of the presented antenna. In terms of improving the antenna parameters, the ground plane in the antenna design is the component that plays the most important role. In most designs, the ground plane is unlimited; however, a significant amount of it was cut out in this design, resulting in improved bandwidth and directivity of the final product.

TABLE 6.1 Designed Parameters of Presented Antenna

Parameters	Value (mm)
Length (L_g)	46.7
Width (W_g)	49.5
Length (L_s)	46.7
Width (W_s)	49.5
Patch Radius (R)	9.2
Square Dimension	5×5
Feed Length (L_F)	2
Feed Width (W_F)	23.75

The specifications of the suggested antenna's design are delineated through key dimensional parameters, each playing a crucial role in determining its overall geometry and performance characteristics. The ground plane, denoted by Length (L_g) and Width (W_g) values of 46.7 and 49.5, respectively, establishes the foundational dimensions of the antenna setup. The patch, characterized by Length (L_s) and Width (W_s) dimensions mirroring those of the ground plane, is designed with precision to optimize its resonance and radiation properties. The inclusion of a Patch Radius (R) measuring 9.2 introduces a curvature element, potentially influencing the antenna's performance. Additionally, a Square Dimension of 5x5 is specified, possibly indicating a square-shaped feature integrated into the patch. The Feed Length (L_F) and Feed Width (W_F) dimensions of 2 and 23.75, respectively, highlight the strategic positioning of the microstrip line feed on the antenna, emphasizing the significance of these dimensions in achieving optimal connectivity and signal transmission. Together, these specified dimensions provide a comprehensive overview of the suggested antenna's physical configuration, setting the stage for detailed analysis and simulation to validate its anticipated performance gains in terms of impedance bandwidth, gain, and efficiency.

6.3 RESULTS AND DISCUSSIONS

In most cases, the ground plane will be far larger than the patch. However, a ground plane that is effectively limitless in size cannot be manufactured. The ground plane should be large. It is recommended that a trade-off be made between the size of the antenna and its performance in order to achieve greater radiation performance and a larger bandwidth.

Figure 6.1 demonstrates a hammer-type stretchable antenna that was constructed using CST software. The triple bandwidths of 20.18%, 17.05%, and 40.09% have been attained, which can be seen in both Table 6.2 and Figure 6.2, both of which make this point quite evident. Figure 6.3 illustrates that the predicted flexible antenna has a maximum directivity of 5.739 dBi at 11.43 GHz. This result can be seen in the figure. Figure 6.4 presents the Smith Chart, a graphical depiction that shows input impedance and may be found in electronic circuits.

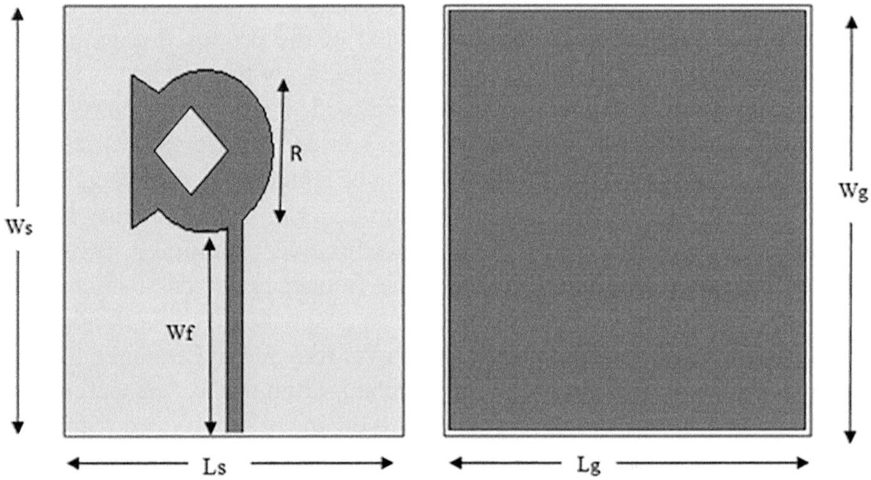

FIGURE 6.1 Structure of anticipated textile antenna.

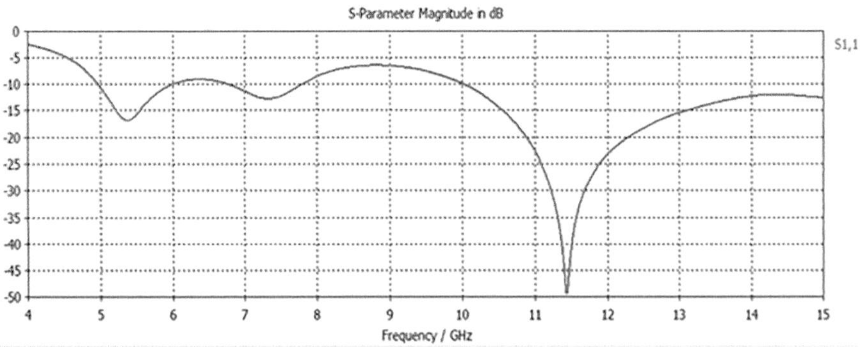

FIGURE 6.2 S_{11} parameter of proposed antenna.

TABLE 6.2 Frequency Range and Bandwidth of Proposed Antenna

Band 1		Band 2		Band 3	
f_1	4.9	f_1	6.75	f_1	9.99
f_2	6	f_2	7.79	f_2	15
BW (%)	20.18	BW (%)	17.05	BW (%)	40.09

The S_{11} parameter of the proposed antenna is a fundamental aspect in evaluating its performance, particularly in terms of impedance matching and signal reflection. S_{11}, also known as the return loss, represents the amount of incident power that is not absorbed by the antenna and is instead reflected back to the source. In the context of the proposed antenna, the S_{11} parameter provides insights into how effectively the antenna is able to transfer power from the transmission line to free space, and conversely, how much power is reflected back to the source. A lower S_{11} value indicates better impedance matching and reduced signal reflection, suggesting that a larger portion of the transmitted signal is radiated efficiently by the antenna. The analysis of the S_{11} parameter involves examining its magnitude and phase characteristics across the frequency range of interest. A well-designed antenna aims to minimize S_{11}, especially within the desired operating frequency band, to optimize power transfer and enhance overall antenna efficiency. Engineers and researchers often use S_{11} measurements to fine-tune antenna designs, ensuring they meet specific performance criteria and standards. By scrutinizing the S_{11} parameter of the proposed antenna, designers can gain valuable insights into its impedance characteristics, enabling them to refine and optimize the antenna for enhanced signal transmission and reception in targeted applications.

The 3-Dimensional Radiation Pattern at 4.587 GHz, as shown in Figure 6.3, serves as a critical representation of how electromagnetic energy is distributed in space by the antenna at this specific frequency. This pattern provides a comprehensive visualization of the antenna's radiation characteristics, illustrating both the magnitude and direction of radiated electromagnetic waves in three-dimensional space. The frequency of 4.587 GHz is a crucial point of analysis, as it corresponds to a specific resonance or operational frequency of the antenna. By examining the 3-D radiation pattern at this frequency, one can gain insights into the antenna's spatial coverage, beamwidth, and radiation efficiency. Variations in radiation intensity across different angles and azimuths become apparent, offering valuable information for applications such as wireless communication or radar systems. The radiation pattern at 4.587 GHz acts as a visual tool for engineers and researchers, aiding in the assessment and optimization of antenna performance at a specific operational frequency, ultimately contributing to the refinement of its design for targeted applications. Figure 6.4 shows the Smith chart of the proposed antenna.

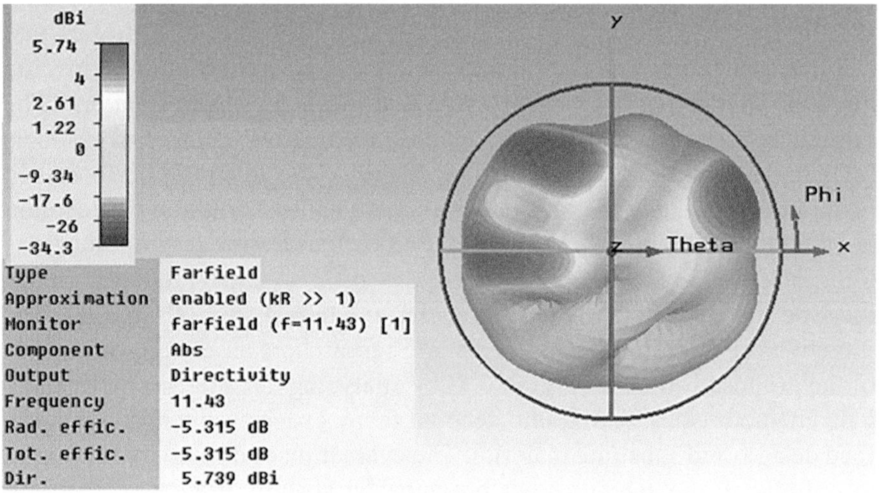

FIGURE 6.3 3-D Radiation pattern at 4.587 GHz.

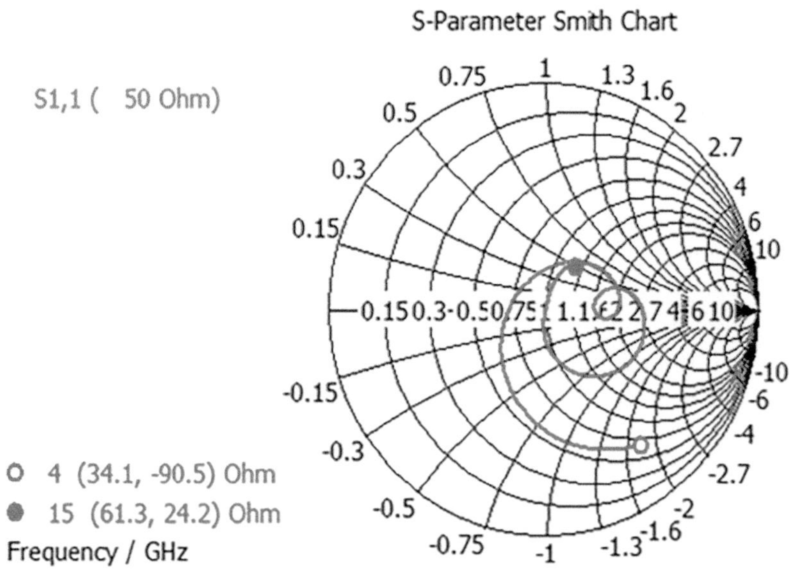

FIGURE 6.4 Smith chart of the proposed antenna.

The directivity of the proposed antenna is a crucial metric that gauges its ability to focus and concentrate electromagnetic radiation in a specific direction, providing a quantitative measure of its directional radiation efficiency. It is essentially a measure of how well the antenna can channel emitted energy into a desired direction compared to an isotropic radiator, which radiates uniformly in all directions. High directivity implies that a significant portion of the radiated energy is concentrated in a specific direction, indicating a more directive and focused radiation pattern. The directivity is typically expressed as a dimensionless ratio or in decibels (dB), and it is a key parameter in understanding the antenna's performance in applications where targeted and precise signal transmission or reception is essential. Achieving high directivity is often desirable in applications like point-to-point communication, radar systems, and satellite communication, where efficient energy transfer and signal strength in a specific direction are paramount. Engineers and researchers assess the directivity of the proposed antenna (Figure 6.5) by analyzing its radiation pattern and gain characteristics, taking into account factors such as antenna geometry, feed design, and substrate material. The evaluation of directivity is integral to optimizing the antenna's performance for specific use cases, ensuring that it meets the requirements of directional communication and sensing applications effectively. Figure 6.6 shows the 2D radiation pattern of the proposed antenna.

6.4 CONCLUSION

The conceptualization and implementation of a novel antenna design are reflected in the development of a unique hammer-type stretchable antenna featuring a square slot. This innovative approach aims to achieve a triple bandwidth encompassing 20.18%, 17.05%, and 40.09% through the incorporation of a strategically placed square slot within the antenna patch. To further enhance the antenna's gain, a technique known as partial ground has been employed, contributing to the realization of a triple wide bandwidth. This approach not only expands the frequency range over which the antenna operates but also improves its efficiency in transmitting and receiving signals. The enhancements in radiation pattern, directivity, and radiation efficiency collectively contribute to the overall performance improvement of the antenna. Additionally, the projected antenna is

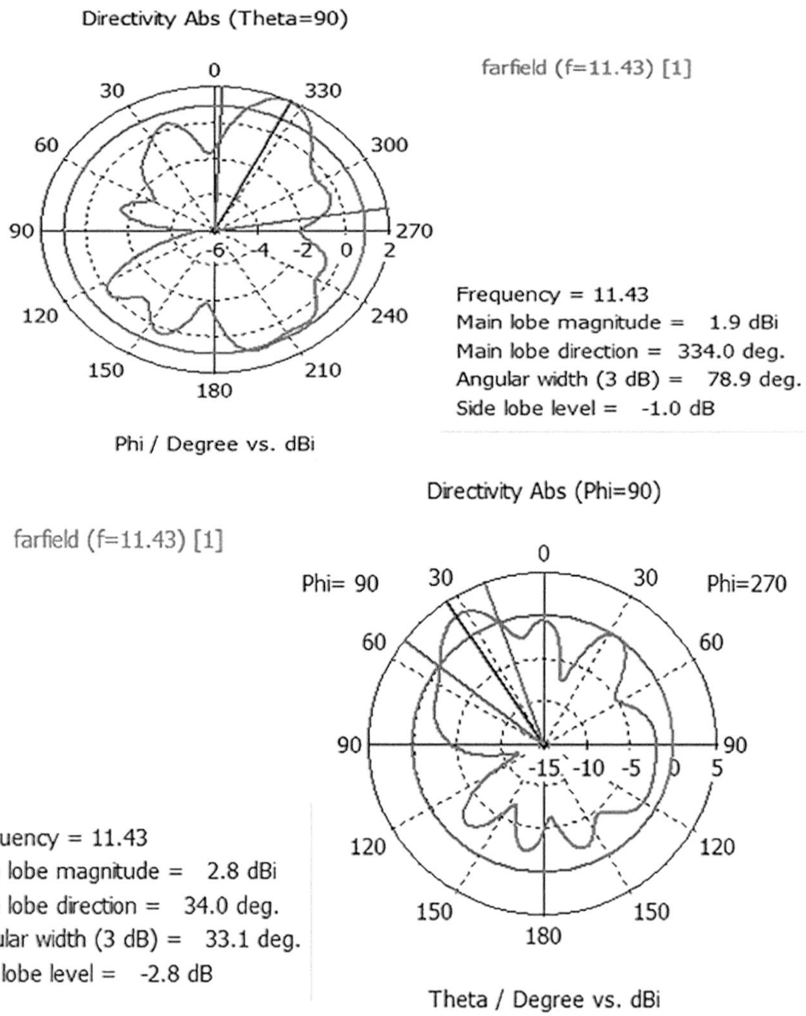

FIGURE 6.5 Directivity of proposed antenna.

anticipated to possess a compact footprint, making it particularly suitable for applications in the IoT realm. The combination of a unique design, triple wide bandwidth, and a compact form factor positions this antenna as an excellent choice for IoT applications, where efficiency, versatility, and a broad operating range are essential considerations.

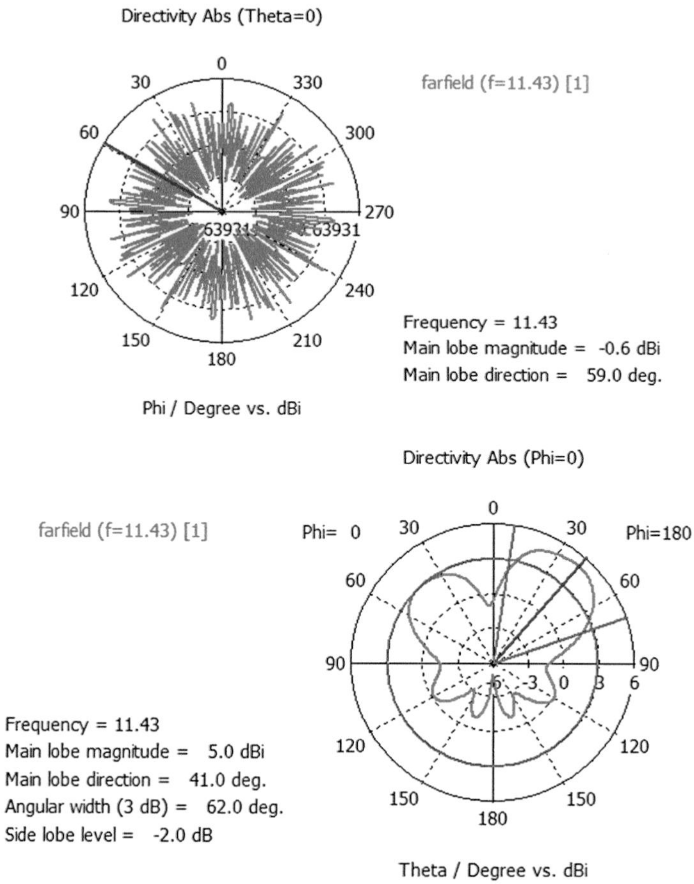

FIGURE 6.6 2-D Radiation pattern at 4.587 GHz.

KEYWORDS

- **hammer-type**
- **Internet of Things**
- **satellite communication**
- **stretchable antenna**
- **triple bandwidth**
- **wireless applications**

REFERENCES

1. Hashmi, S. N. A., Singh, V. K., Dwivedi, R. K., Pathak, R. S., & Tiwari, R. (2022). Tri pentagon slotted antenna using jeans material for WLAN/Hiper LAN applications. In *2022 International Conference on Advancements in Smart, Secure and Intelligent Computing (ASSIC)* (pp. 1–4). IEEE.
2. Kakkar, S., Kamal, T. S., & Singh, A. P. (2019). On the design and analysis of I-shaped fractal antenna for emergency management. *IETE Journal of Research, 65*(1), 104–113.
3. Saxena, A., & Singh, V. K. (2022). Design of compact array antenna and its effect on the human brain. *Wireless Personal Communications, 125*(1), 637–647.
4. Yadav, A., Singh, V. K., Yadav, P., Beliya, A. K., Bhoi, A. K., & Barsocchi, P. (2020). Design of circularly polarized triple-band wearable textile antenna with safe low SAR for human health. *Electronics, 9*(9), 1366.
5. Singhal, A. K., Beniwal, N. S., Almutairi, K., Arockia Dhanraj, J., Mostafaeipour, A., Issakhov, A., & Singh, V. K. (2021). Experimental analysis of modified DC-P&O technique with arm controller for a stand-alone 40 W PV system. *Energies, 14*(19), 6169.
6. Bangari, N., Singh, V. K., & Sharma, V. K. (2020). Experimental investigation of thin-film solar cells as a wearable power source. *Energy Sources, Part A: Recovery, Utilization, and Environmental Effects, 46*(1), 9341–9361.
7. Yadav, A., Kumar Singh, V., Kumar Bhoi, A., Marques, G., Garcia-Zapirain, B., & de la Torre Díez, I. (2020). Wireless body area networks: UWB wearable textile antenna for telemedicine and mobile health systems. *Micromachines, 11*(6), 558.
8. Singh, V. K., Lala, A., & Singh, A. K. (2020). Novel inset feed circular slotted microstrip antenna using multilayer feed-forward back-propagation and radial basis function neural network. *National Academy Science Letters, 43*, 343–345.
9. Yadav, A., Singh, V. K., & Mohan, H. (2019). Design of a U-shaped circularly polarized wearable antenna with DGS on a fabric substrate for WLAN and C-band applications. *Journal of Computational Electronics, 18*(3), 1103–1109.
10. Tripathi, S., Tiwari, R., Shrivastava, A. K., Singh, V. K., Dubey, N., & Dubey, V. (2018). A review reports on rare earth activated AZrO3 (A= Ba, Ca, Sr) phosphors for display and sensing applications. *Optik, 157*, 365–381.
11. Singh, V. K., Dhupkariya, S., & Bangari, N. (2017). Wearable ultra-wide dual-band flexible textile antenna for WiMax/WLAN application. *Wireless Personal Communications, 95*(2), 1075–1086.
12. Singh, N., Singh, A. K., & Singh, V. K. (2015). Design & performance of wearable ultra-wideband textile antenna for medical applications. *Open Engineering, 57*(7), 117–123.
13. Singh, R., Singh, V. K., & Khanna, P. (2018). A compact CPW-fed defected ground microstrip antenna for Ku-band application. In *Advances in Electronics, Communication and Computing: ETAEERE-2016* (pp. 231–237). Springer Singapore.
14. Singh, N. K., Singh, V. K., Saxena, A., Bhoi, A. K., Garg, A., & Sherpa, K. S. (2018). A compact slotted textile patch antenna for ultra-wideband application. In *Advances*

in Electronics, Communication and Computing: ETAEERE-2016 (pp. 53–59). Springer Singapore.

15. Singhal, S., Goel, T., & Singh, A. K. (2017). Hexagonal tree shaped ultra-wideband fractal antenna. *International Journal of Electronics Letters, 5*(3), 335–348.
16. Singh, V. K., Ali, Z., Singh, A. K., & Ayub, S. (2014). A wideband compact microstrip antenna for DCS/PCS/WLAN applications. In *Intelligent Computing, Networking, and Informatics: Proceedings of the International Conference on Advanced Computing, Networking, and Informatics, India, June 2013* (pp. 1107–1113). Springer India.

CHAPTER 7

5G Communication and IoT

PARMINDER KAUR,[1] MANISH SHARMA,[1,*] and MANOJ KAPIL[2]

[1]*Chitkara University Institute of Engineering and Technology, Chitkara University, Punjab, India*

[2]*Department of Computer Science, Houston Tillotson University, Texas, USA*

*Corresponding author

ABSTRACT

5G is getting attention in almost every field of life owing to its capacity for higher data rates, lower latency, and higher bandwidth as well. 4G has already been implemented for wireless communication but couldn't fulfill the inflating demands of modern mobile users such as cloud services, video streaming, and social applications. To meet these requirements, ITU has set another standard called 5G, which includes several frequency bands from sub-6 GHz to millimeter wave frequency (30 GHz–300 GHz) bands. The millimeter-wave spectrum also finds its application in IoT for the virtualization of devices, services, and things connected in the network. The chapter aims at 5G communication, infrastructure, architecture, specifications, usage scenarios, 5G components, and technologies, along with the application of machine and deep learning in IoT using energy-efficient antennas for 5G.

7.1 INTRODUCTION

India, a developing country, is progressing in every field of the modern era. The rising demands for data and compact devices have revolutionized the field of technology as well as the spectrum. The majority of applications today are cloud-based, necessitating constant, low latency, and high-speed

Wireless Technology: Advances and Innovations for Smart IoT Applications. Rajesh Kumar Dwivedi, Vinod Kumar Singh, and Raghav Dwivedi (Eds.)

DOI: 10.1201/9781998511495-7

data. Yet, while 4G has been implemented for most applications, it still does not satisfy the demands for higher bandwidth and data rates. Because of its constrained bandwidth and growing user base, 4G Long Term Evolution (LTE) has currently hit its data support saturation limit. According to the CISCO Visual Network Index (VNI) 2017–2022, the world is entering an era of zettabytes, and global IP traffic is likely to increase threefold from 2017–2022. Mobile devices are predicted to grow to 13.1 billion by 2023. Therefore, there will also be a significant increase in the number of internet users, devices and networks, broadband speeds, video streaming, and so forth [1]. As 4G, notwithstanding its limitations, faces these challenges, the International Telecommunication Union (ITU) has set a new standard for wireless communication called the 5G mm-wave spectrum, which will provide a data rate of about 10 Gbps, a latency of less than 1 ms, and high capacity with a broad spectrum in the 30 GHz–300 GHz band [2]. The introduction of mm-wave for 5G will increase the potential for greater data throughput because of its broad bandwidth and speed. Wireless communication has proven to be a boon in the field of communication as it eliminated the need for connecting additional wires to enhance the capacity of the system. However, it also has some limitations, such as low connectivity speed, lower bandwidth, and congestion in channels. To address these challenges, technological transformation has taken place from 1G–5G and beyond, which has been described in Section 7.1.

Figure 7.1 shows the evolution of wireless technology from 1G–6G. Digital Enhanced Cordless Communication (DECT) and Advanced Mobile Phone Services (AMPS) were introduced as the standards for analog signal voice services in the 800 MHz–900 MHz spectrum in 1G, with a data capacity of 2 Kbps in the 1980s. The first 1G systems included Nordisk Mobile Telephony (NMT), AMPS, and Cellular Digital Packet Data (CDPD) technologies and were analog frequency-modulated systems. These standards were eventually replaced by Information and Communication Technology (ICT) as well as by Global System for Mobile Communications (GSM) in the 1990s for voice and data transfer at less than 2 GHz frequencies. 2G found its foundation with digital technologies such as Time Division Multiple Access (TDMA) and Code Division Multiple Access (CDMA) as well [3, 4]. The third generation, 3G, came up with the Universal Mobile Telecommunications System (UMTS) developed by 3GPP in IMT-2000. Many technologies such as W-CDMA, UMTS, DECT, and GSM-EDGE were supported by 3G for voice and internet communication. LTE and Worldwide Interoperability for Microwave Access (WiMAX) were launched as upgrades to 3G standards in 2010 to increase the data rate and capacity of 3G. IMT-Advanced introduced 4G as a network providing data transfer at less than 6 GHz for internet technology along with other applications such as Wi-Fi and WiMAX. It provided a data capacity of up to 1 Gbps [5, 6]. In 2020, verticals were introduced at less than 100 GHz by 3GPP 5G NR, enabling 5G to be flexibly deployed and scaled with a data capacity of greater than 1 Gbps [7]. All the existing generations have focused on human communications through voice, data, and the internet, whereas 5G networks are focusing on industrial communications for the global digitization of the economy by enhancing various parameters of networks such as low latency, throughput, reliability, and so forth, for <100 GHz frequencies, as stated in IMT-2020 [2]. Last but not least, the sixth generation (6G) is said to be introduced in 2030 in THz frequencies (<1 THz) as specified in IMT-2030 by ITU-R, with the fusion of machine learning and artificial intelligence as well.

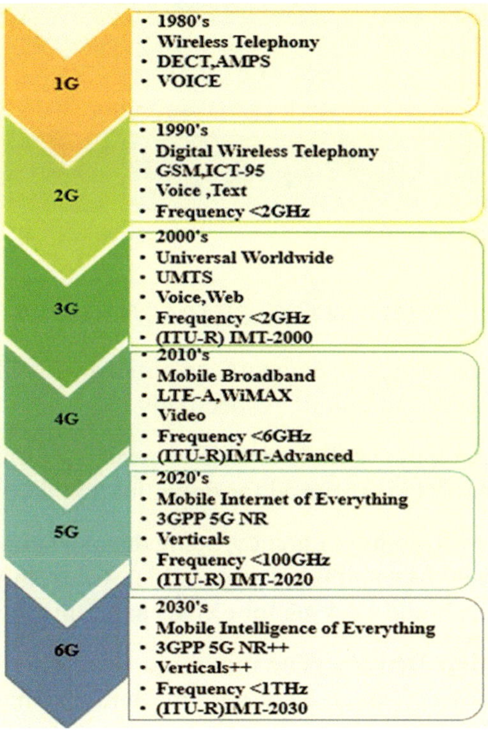

FIGURE 7.1 Cellular evolution.

7.1.1 FIFTH GENERATION (5G) INTRODUCTION

5G has made the integration of several technologies, networks, and applications possible to provide various value-added services like smart vehicles, video streaming, virtual reality, and cloud services. The ITU has set various characteristics that must be met by 5G devices, of which the important ones are represented in Figure 7.2 and are described below [2].

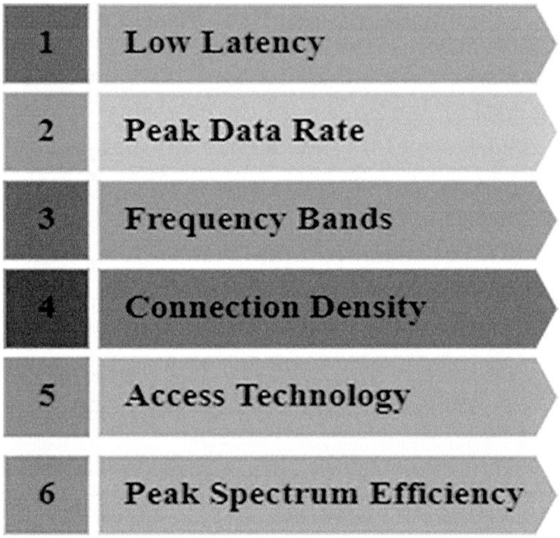

FIGURE 7.2 5G characteristics.

7.1.2 FIFTH GENERATION (5G) CHARACTERISTICS

1. **Data Rate:** The peak data rates required are 20 Gbps and 10 Gbps for downlink and uplink, respectively.
2. **Peak Spectrum Efficiency:** The peak spectrum efficiency of 30/15 bits/s/Hz is required for downlink/uplink data.
3. **Ultra-Low Latency:** Latency, being the most important characteristic, should be as minimal as possible. According to the standards of ITU, it should be 4 ms for eMBB and 1 ms for uRLLC.
4. **Connection Density:** The minimum requirement for connection density is 1 million devices per km^2, with a connection reliability of $1–10^5$ success probability.

5. **Access Technology:** 5G employs the application of beam division multiple access (BDMA) to increase the capacity of the network by generating highly directive beams. It also uses filter bank multicarrier (FBMC) techniques for better spectral efficiency.
6. **Frequency Bands:** 5G utilizes two spectrums classified as FR1 for frequencies below 6 GHz and FR2 for mm-wave frequencies greater than 6 GHz, such as 26/28/37/39/47/60 GHz [8].

7.1.3 FIFTH GENERATION (5G) USAGE SCENARIOS

Enhanced Mobile Broadband Services (eMBB), Ultra-Reliable Low Latency Communication (uRLLC), and Massive Machine Type Communication (mMTC), which have specific technical requirements to be met in accordance with IMT-2020 objectives, are the three 5G scenarios defined by 3GPP (The 3rd Generation Partnership Project) and are represented in Figure 7.3. Bandwidth, being the general scenario, should be a minimum of 100 MHz and up to 1 GHz for frequency bands greater than 6 GHz. eMBB is to be employed for high-speed indoor and outdoor connections, with data rates of 20 Gbps for indoor and 2 Gbps for outdoor, respectively. mMTC supports IoT, and more than 10,000 devices can be supported by a single base station. uRLLC accounts for a low packet loss of 1 in 10,000, with the lowest latency of 1 ms for remote and safety applications [2, 9].

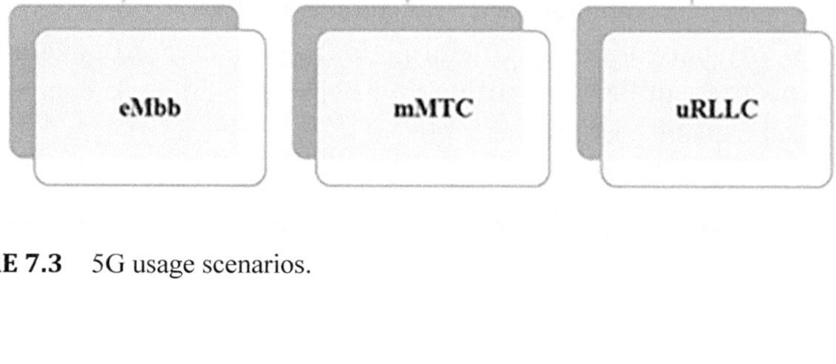

FIGURE 7.3 5G usage scenarios.

7.2 FIFTH GENERATION (5G) FREQUENCY BANDS

5G spectrum characterized by the FCC consists of three bands: Lower Band, Mid Band, and Higher Band. All these bands have certain frequency ranges, which are shown in Figure 7.4 [2, 10]. The lower band, also known as the sub-1 GHz band, ranges from 600 MHz–900 MHz. The mid band consists of 2.5 GHz, 3.5 GHz, along with the 3.7 GHz–4.2 GHz frequency spectrum. The higher band, also known as the millimeter band, comprises frequencies from 24 GHz–47 GHz, with wavelengths of 1 mm–0 mm.

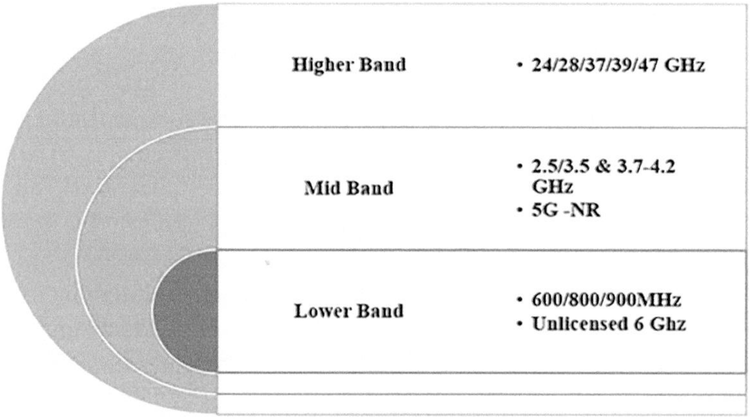

FIGURE 7.4 5G frequency bands.

7.3 TECHNOLOGY ENABLERS AND ITS COMPONENTS FOR FIFTH GENERATION (5G)

There are certain technology enablers that contribute to the implementation and challenges of 5G. Some of them have been listed in Figure 7.5 and are described as follows [11–17]:

1. **Device-to-Device Communication (D2D):** There is direct communication between the devices, without the involvement of any network infrastructure, aiming at increased coverage and spectrum utilization, offloading backhaul, fallback connectivity, and capacity per area as well.
2. **Massive Machine Communication (MMC):** With varying data rates, prices, and latency, it offers scalable connection solutions for billions of network-enabled devices.

3. **Moving Networks (MNs):** Populations that are a subset of cooperatively moving communication devices are covered by moving networks to a greater extent. A collection of MN nodes that connect with other nodes that may be fixed, mobile, or situated inside or outside the moving entity makes up a moving network.
4. **Ultra-Dense Networks (UDNs):** These networks provide dense infrastructure to meet the requirement for heavy throughput, thereby supporting increased radio connection capacity, energy efficiency, and spectrum exploration. However, due to the increased number of interior and outdoor access points, there are more issues related to mobility, cost, and interference.
5. **Ultra-Reliable Communication (URC):** High levels of availability and dependability are made possible through this communication.
6. **Massive Multiple Input Multiple Output (MIMO):** Massive MIMO involves transmitters and receivers outfitted with a high number of antenna components that may be co-located or distributed to improve spectral as well as energy efficiency.
7. **Millimeter-Wave Technology (MMT):** The 3 GHz–300 GHz spectrum has led to an increase in bandwidth as well as data rates up to Gbps with low latency.
8. **Cloud Technology:** It enables on-demand access to computing resources from a common or shared pool. The technology uses the internet and a central remote server or content provider for using applications without installation.
9. **Green Communication:** The designs of 5G devices or systems should consume minimal energy to reduce carbon dioxide emissions and contribute to green wireless communication.
10. **Network Slicing:** Network slicing divides the network space into chunks or logical self-contained networks that have the resources to run the network for the user's specific requirements. Once the network is tuned by the network operator, the user can have full access to 5G infrastructure vertical layers, which can also pose a threat to the logical network.
11. **Non-Orthogonal Multiple Access (NOMA):** It assigns different powers to a pair of users sharing the same bandwidth resource.

When compared to orthogonal multiple access strategies such as TDMA, FDMA, CDMA, and OFDMA, NOMA is known to increase the throughput of cell edge users.

12. **Full Duplex Radio:** Full duplex communication supports two-way communication simultaneously. It doubles spectrum efficiency, improves network efficiency, and helps in reducing end-to-end packet delay.
13. **Architecture (Arch):** There must be a novel architecture that provides a consistent integration of centralized and decentralized approaches.

FIGURE 7.5 5G technology enablers.

7.4 TECHNOLOGY COMPONENTS

In the realm of telecommunications, 5G technology represents a cutting-edge paradigm characterized by the integration of four fundamental components that collectively redefine the landscape of wireless communication. First and foremost, the foundation of 5G is laid upon robust and efficient radio links, embodying the wireless connections that facilitate seamless communication between devices. This encompasses the utilization of radio frequency signals to achieve unparalleled data transfer rates and minimized latency, setting 5G apart from its predecessors. The second key component, multi-antenna technology, introduces a sophisticated approach known as MIMO, employing multiple antennas at both transmitting and receiving ends. This strategic implementation enhances signal quality, augments data capacity, and optimizes performance, particularly in challenging environments. Complementing these advancements, 5G embraces the concept of multilayer networks, diversifying the spectrum bands used for communication. This entails the deployment of a spectrum ranging from low-band to high-band (mmWave), allowing 5G to balance coverage and capacity demands efficiently. Finally, the fourth component, spectrum usage, emphasizes the judicious allocation of frequencies for wireless communication. By exploring new spectrum bands, including high-frequency millimeter waves, 5G maximizes data speeds and capacity, catering to the escalating demand for wireless data. Figure 7.6 likely illustrates the intricate interplay and synergy among these components, providing a visual representation of how they collectively contribute to the transformative capabilities of 5G technology are follows:

1. **Radio Links:** Radio links must be flexible to support different scenarios and different users. They must be scalable for ultra-dense networks to encounter different types of signals in different bands. Moreover, they should be optimized for machine-type as well as human-centric communications. Therefore, they should be scalable, flexible, and optimizable for all types of communications, as shown in Figure 7.7.

2. **Multi-Node/Multi-Antenna:** This component plays an important role in communication as it involves MIMO terminals for higher capacity, spectral efficiency, and data rates. These also provide improved link reliability, energy efficiency, and coverage. Various ways to implement multi-antennas include arrays, massive MIMO

in mm-wave bands, multi-hop technologies, and inter-node coordinate schemes, as depicted in Figure 7.8, which has been redrawn from [14].

3. **Network Dimension:** It comprises approaches for effective interference control for deployment in complicated heterogeneous contexts, as well as demand, traffic, and mobility management.
4. **Spectrum Usage:** It is divided into two parts: the extended band of operation and the new spectrum addressing the needs of each usage scenario of 5G specified by IMT-2020.

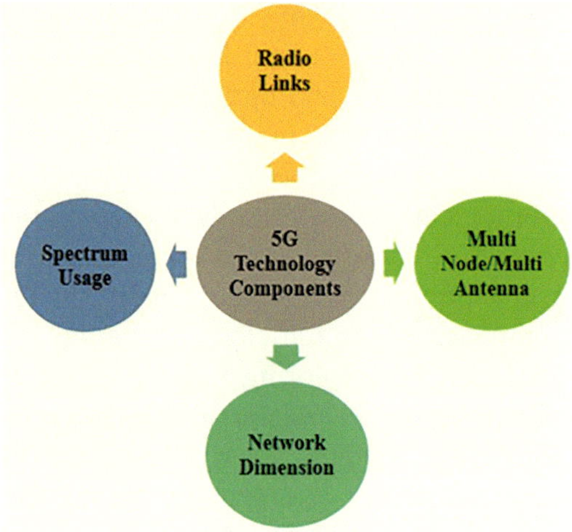

FIGURE 7.6 5G technology components.

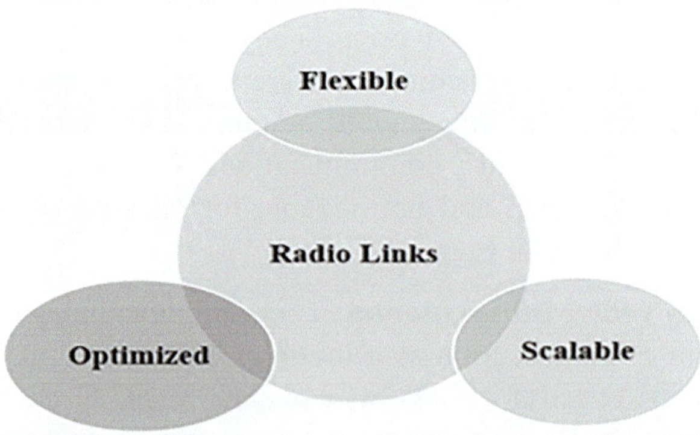

FIGURE 7.7 Radio links characteristics.

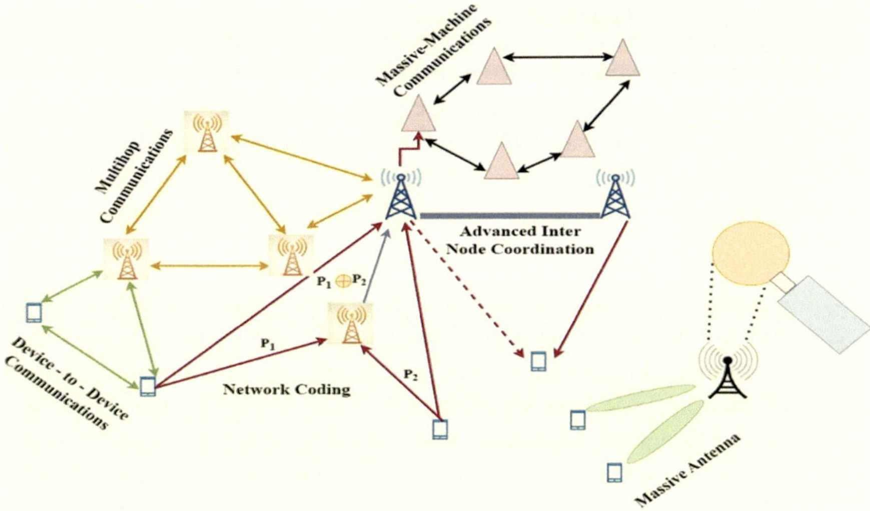

FIGURE 7.8 Multi-Node transmission.

7.5 ARCHITECTURES IN FIFTH GENERATION (5G) AND IOT

5G architecture is the pillar for IoT applications, and it has to be redesigned to provide connectivity with speeds of more than 10 Gbps among various IoT applications. The architecture components for 5G are described as follows and are depicted in Figure 7.9.

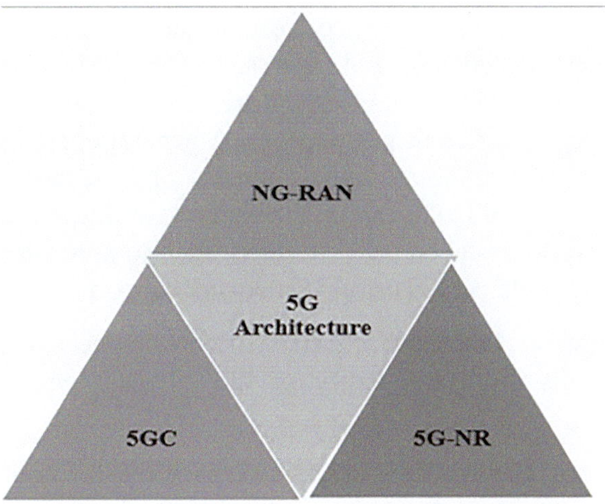

FIGURE 7.9 5G-architecture.

7.5.1 5G ARCHITECTURE

1. **Next Generation Radio Access Network (NG-RAN):** It consists of a base station providing access to 5G-enabled user equipment, 5G-Core (5GC), and 4G-Evolved Packet Core (EPC) by using the 5G New Radio air interface. It also provides the connection of the base station to 5GC by accessing 5G-enabled UE over 4G radio.
2. **5GC:** It indicates 5G-Core service based on various interconnected network functions. It comprises control and user planes, which further perform their specific functions. Access and Mobility Management Function (AMF) and Session Management Function (SMF) are handled by the 5GC control plane, while User Plane Function is handled by the 5G user plane (UPF). For greater performance, there is a system called 5GC Control and User Plane Separation (CUPS) that distributes user plane tasks to sites closest to the users while centralizing the control plane tasks. Interfaces are also provided for the base stations and the servers.
3. **5G-NR:** 5G New Radio has made the spectrum flexible, scalable, and efficient for users. Spectrum bands include the sub-6 GHz band and the mm-wave band of 30 GHz–300 GHz.

7.5.2 IOT ARCHITECTURE

Today's IoT applications enable comfortable and smarter livelihoods for everyone by interconnecting several smart devices around and across buildings, cities, and industries. Many architectures are being proposed by researchers for IoT, considering different perspectives. A general approach toward IoT architecture is shown in Figure 7.10. It consists of three basic layers for the general requirements of IoT: Perception, Network, and Application layers, as described below. Seven-layered architectures are also used where smart businesses are being implemented using 5G network support are:

1. **Perception Layer:** It consists of the devices that collect data physically from the entity or the environment; For example, sensors.
2. **Network Layer:** This layer is responsible for connections with other devices and services, along with the processing of data received from sensors in the perception layer.

3. **Application Layer:** It decides the applications to be implemented for IoT, including the services to be provided to the users.

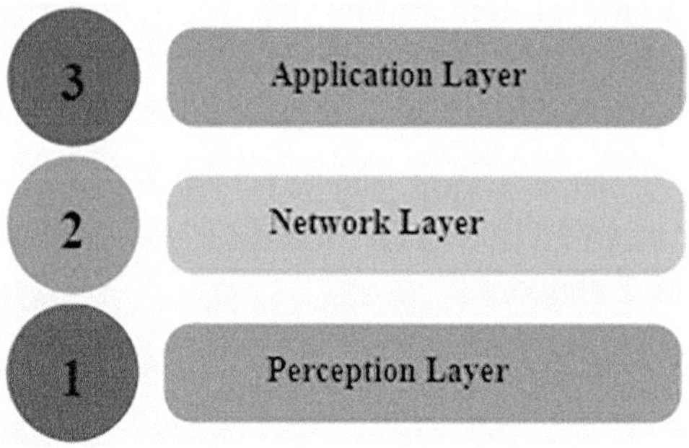

FIGURE 7.10 Three layered IoT-architecture.

7.6 FIFTH GENERATION (5G) AND INTERNET OF THINGS (IOT)

5G and IoT are interdependent technologies. Where 5G offers a broad spectrum IoT provides the connectivity of devices. Antenna being the most important part of any wireless network serves as an interface between the IoT device and the network. It contributes in achieving high gain, good signal propagation in mobile IoT device with improved signal levels. Literature has been surveyed which features both technologies together for various applications and is discussed below:

On-Body Communication: Two removable Finger-Nail antennas at 15 and 28 GHz has been proposed [17] for sensing the data in 5G spectrum for IoT applications as shown in Figure 7.11(a) and 7.11(b).

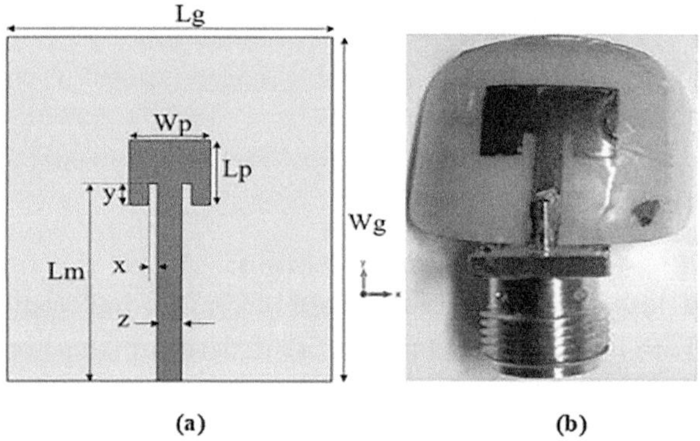

FIGURE 7.11 (a) Antenna design (b) Fabricated antenna.

Source: Reprinted from Ref. [17]. https://creativecommons.org/licenses/by/4.0/

1. **Manufacturing Process:** As a huge amount of data is generated during the manufacturing process, to achieve real-time transmission and processing of data at high data rates, coverage, reliability, and low latency, a cyber-physical manufacturing system is proposed in a 5G network for smart manufacturing in Industrial IoT [18].
2. **Communication Using Sensors:** A multi-layer Yagi antenna operating at 24 GHz is presented in [19] to operate in the mm-wave band for IoT sensors.
3. **Railways Communication:** An antenna with a 28 GHz resonant frequency in the 5G spectrum is reported in [20] for railway communications. It is designed to be mounted on the roof of the train for communication in the front/back directions as well as with the base stations installed on the sides of the tracks. The data is then transmitted through the server to the connected or monitoring devices, as shown in Figure 7.12, being redrawn from [20].

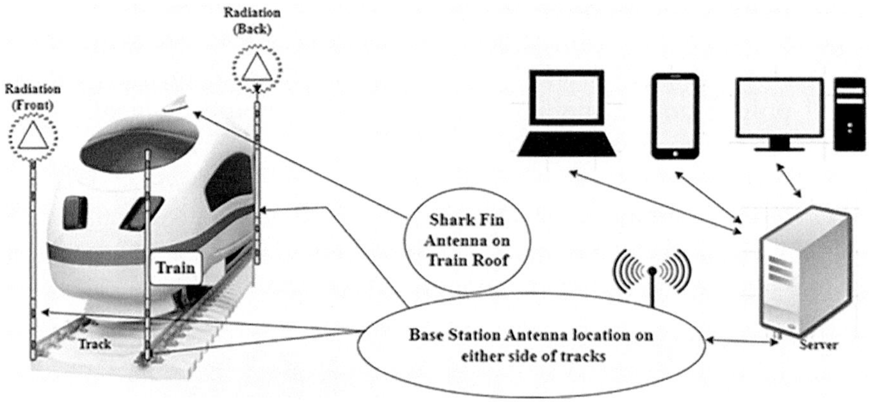

FIGURE 7.12 Railways communication in IoT.
Source: Adapted from Ref. [20]

5. **Textile Antenna Communication:** A textile antenna at a frequency of 26 GHz is presented in [21] for IoT applications. The Electromagnetic Bandgap (EBG) structure placed on a polyester fabric has improved the performance of the antenna in terms of gain, energy efficiency, and impedance matching. The simulated structure is shown in Figure 7.13(a), and the regenerated scattering parameter (S_{11}) is presented in Figure 7.13(b) referenced in [21].

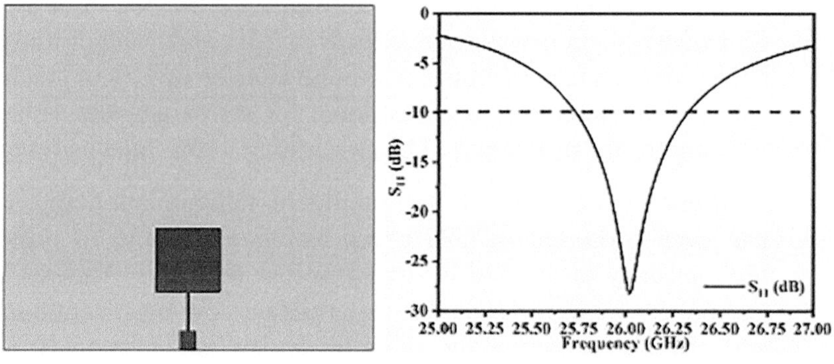

(a) (b)

FIGURE 7.13 (a) Textile antenna structure. (b) S-Parameter of textile antenna.

7.7 MACHINE LEARNING (ML) AND FIFTH GENERATION (5G)

1. **Detection of Unmanned Aerial Vehicles (UAV):** A method for the detection, identification, and positioning of UAVs by radar is proposed in [22] using high-resolution range profile technology. UAVs are detected and positioned using the mm-wave spectrum of the 5G network without the involvement of GPS, as shown in Figure 7.14, redrawn from [22].

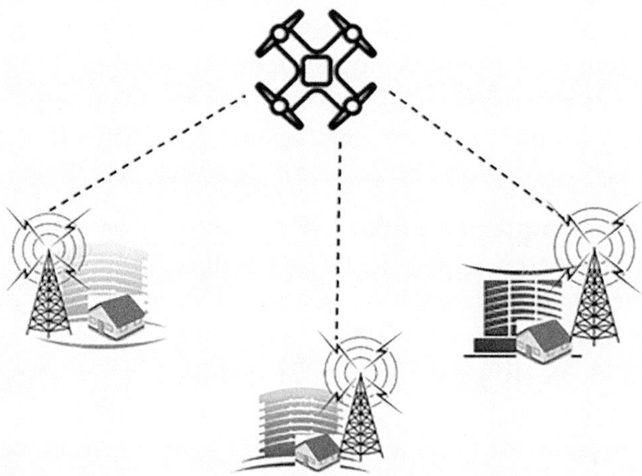

FIGURE 7.14 UAV detection with 5G in IoT.

Source: Adapted from Ref. [22]

2. **Beam Forming:** An adaptive beam-forming method is designed for MIMO-5G mm-wave networks using ML, and an algorithm has been proposed [23] to reduce noise and interference. Four machine learning methods have been explored for six parameters of beam forming for the estimation of beam-forming in machine learning.
3. **Authentication Method:** An intelligent authentication approach has been proposed in [24] using machine learning to provide high security in 5G and beyond wireless networks. Various ML techniques are used for designing a reliable, low-cost, continuous, and situation-aware authentication device under unpredictable network conditions.
4. **Security:** A physical-layer authentication algorithm is reported in [25] using support vector machine (SVM) to identify security threats in 5G wireless communication.

7.8 DEEP LEARNING (DL) AND FIFTH GENERATION (5G)

1. **Path Forecasting AND Altitude Assessment:** A new deep learning technique (DL) is presented in [26] for path forecasting and altitude assessment; it utilizes multiple-beam-forming technology to improve the data rate of UAVs for 5G-IoT networks.
2. **Smart Beam-forming System:** In [27], deep learning is used for an intelligent beamforming system. An ideal antenna schematic with an extension to MIMO is proposed for a large number of cases without solving mathematically difficult optimization issues.
3. **Testbed Implementation:** C-RAN and IoT networks are created in [28] for an experimental 5G testbed, which improves both vertical and horizontal localization. A deep-learning-based cooperative architecture (DELTA), built on a three-dimensional, multilayered identification broadcast database, is recommended for its implementation. Two-dimensional coordinates are calculated, followed by the prediction of the 3D mobile station location.
4. **Blockchain-based Security Approach:** In [29], a security framework has been proposed for four hierarchical layers such as

cloud, fog, edge, and user to support intelligent data and security operations by using deep learning and blockchain. Feasibility is being checked by object detection tasks, taking latency, accuracy, and security into consideration for 5G-enabled IoT networks.

5. **Handover Mechanism Implementation:** In [28], a quick and dependable LSTM algorithm based on deep learning is presented as a handover mechanism. To construct the user's profile, movement patterns are observed. The accuracy rate of detected attackers rises as the volume of authentication requests grows, and the likelihood of false positives decreases.

7.9 CONCLUSION

The chapter concludes that 5G has great potential for providing high data rates and ultra-high latency in almost all vertical sectors such as communication, healthcare, agriculture, military, and security, and so forth. It has numerous technology enablers that are to be used for the virtualization of devices as well as services in connected networks. It is also concluded that 5G can be implemented by incorporating machine and deep learning algorithms for various applications in IoT, such as beam forming, authentication, security, railway communication, data transmission, path finding, location prediction, and so forth. The chapter also focuses on the effective interdependence of technologies, which will help in achieving the virtualization of everything to be available everywhere.

KEYWORDS

- **5G technology**
- **data transfer**
- **higher band**
- **lower band**
- **machine learning**
- **massive MIMO**
- **millimeter-wave**

REFERENCES

1. Khan, J., Ullah, S., Ali, U., Tahir, F. A., Peter, I., & Matekovits, L. (2022). Design of a millimeter-wave MIMO antenna array for 5G communication terminals. *Sensors, 22*(7), 2768.
2. Kumar, S., Dixit, A. S., Malekar, R. R., Raut, H. D., & Shevada, L. K. (2020). Fifth generation antennas: A comprehensive review of design and performance enhancement techniques. *IEEE Access, 8*, 163568–163593.
3. Shakir, M., Aslam, S., Sarwar, M. U., Adnan, M., & Khan, M. R. (2021). Performance evaluation and design of 5G communication-based millimeter wave antenna. *EURASIP Journal on Wireless Communications and Networking, 2021*, 1–10.
4. Akyildiz, I. F., Gutierrez-Estevez, D. M., & Reyes, E. C. (2010). The evolution to 4G cellular systems: LTE-Advanced. *Physical Communication, 3*(4), 217–244.
5. Mondal, S., Sinha, A., & Routh, J. (2015). A survey on evolution of wireless generations 0G to 7G. *International Journal of Advance Research in Science and Engineering (IJARSE), 1*(2), 5–10.
6. Juneja, S., Pratap, R., & Sharma, R. (2021). Semiconductor technologies for 5G implementation at millimeter wave frequencies–Design challenges and current state of work. *Engineering Science and Technology, an International Journal, 24*(1), 205–217.
7. Nowak, T. W., Sepczuk, M., Kotulski, Z., Niewolski, W., Artych, R., Bocianiak, K., & Wary, J. P. (2021). Verticals in 5G MEC-use cases and security challenges. *IEEE Access, 9*, 87251–87298.
8. Juneja, S., Pratap, R., & Sharma, R. (2021). Design considerations for implementation of planar antennas for millimeter wave (mmW) 5G network: A review. *Circuit World.*
9. Ikram, M., Sultan, K., Lateef, M. F., & Alqadami, A. S. (2022). A road towards 6G communication—A review of 5G antennas, arrays, and wearable devices. *Electronics, 11*(1), 169.
10. Dogra, A., Jha, R. K., & Jain, S. (2020). A survey on beyond 5G network with the advent of 6G: Architecture and emerging technologies. *IEEE Access, 9*, 67512–67547.
11. Wang, C. X., Haider, F., Gao, X., You, X. H., Yang, Y., Yuan, D., & Hepsaydir, E. (2014). Cellular architecture and key technologies for 5G wireless communication networks. *IEEE Communications Magazine, 52*(2), 122–130.
12. Ma, Z., Zhang, Z., Ding, Z., Fan, P., & Li, H. (2015). Key techniques for 5G wireless communications: Network architecture, physical layer, and MAC layer perspectives. *Science China Information Sciences, 58*(4), 1–20.
13. Patil, G. R., & Wankhade, P. S. (2012). 5G wireless technology. *World, 3*.
14. Osseiran, A., Boccardi, F., Braun, V., Kusume, K., Marsch, P., Maternia, M., & Fallgren, M. (2014). Scenarios for 5G mobile and wireless communications: The vision of the METIS project. *IEEE Communications Magazine, 52*(5), 26–35.
15. Shakhovska, N., Vysotska, V., & Chyrun, L. (2017). Intelligent systems design of distance learning realization for modern youth promotion and involvement in independent scientific researches. In *Advances in Intelligent Systems and Computing: Selected Papers from the International Conference on Computer Science and*

Information Technologies, CSIT 2016, September 6–10, Lviv, Ukraine (pp. 175–198). Springer International Publishing.
16. Gupta, A., & Jha, R. K. (2015). A survey of 5G network: Architecture and emerging technologies. *IEEE Access, 3*, 1206–1232.
17. Njogu, P., Sanz-Izquierdo, B., Elibiary, A., Jun, S. Y., Chen, Z., & Bird, D. (2020). 3D printed fingernail antennas for 5G applications. *IEEE Access, 8*, 228711–228719.
18. Cheng, J., Chen, W., Tao, F., & Lin, C. L. (2018). Industrial IoT in 5G environment towards smart manufacturing. *Journal of Industrial Information Integration, 10*, 10–19.
19. Ramos, A., Varum, T., & Matos, J. N. (2018). Compact multilayer Yagi-Uda based antenna for IoT/5G sensors. *Sensors, 18*(9), 2914.
20. Arya, A. K., Kim, S., Ko, K., & Kim, S. (2021). Antenna for IoT-based future advanced (5G) railway communication with end-fire radiation. *IEEE Internet of Things Journal, 9*(9), 7036–7042.
21. Wissem, E. M., Sfar, I., Osman, L., & Ribero, J. M. (2021). A textile EBG-based antenna for future 5G-IoT millimeter-wave applications. *Electronics, 10*(2), 154.
22. Zhao, J., Fu, X., Yang, Z., & Xu, F. (2019). Radar-assisted UAV detection and identification based on 5G in the Internet of Things. *Wireless Communications and Mobile Computing, 2019(4), 1–12.*
23. Chopra, S. (2021). Design analysis of adaptive beamforming in a MIMO-millimeter wave 5G heterogeneous wireless network using machine learning models. DOI: 10.21203/rs.3.rs-312958/v1.
24. Fang, H., Wang, X., & Tomasin, S. (2019). Machine learning for intelligent authentication in 5G and beyond wireless networks. *IEEE Wireless Communications, 26*(5), 55–61.
25. Sumathy, S., Revathy, M., & Manikandan, R. (2023). Improving the state of materials in cybersecurity attack detection in 5G wireless systems using machine learning. *Materials Today: Proceedings, 81*, 700–707.
26. Zhu, X., Qi, F., & Feng, Y. (2020). Deep-learning-based multiple beamforming for 5G UAV IoT networks. *IEEE Network, 34*(5), 32–38.
27. Maksymyuk, T., Gazda, J., Yaremko, O., & Nevinskiy, D. (2018). Deep learning-based massive MIMO beamforming for 5G mobile network. In *2018 IEEE 4th International Symposium on Wireless Systems within the International Conferences on Intelligent Data Acquisition and Advanced Computing Systems (IDAACS-SWS)* (pp. 241–244). IEEE.
28. El Boudani, B., Kanaris, L., Kokkinis, A., Kyriacou, M., Chrysoulas, C., Stavrou, S., & Dagiuklas, T. (2020). Implementing deep learning techniques in 5G IoT networks for 3D indoor positioning: DELTA (DeEp Learning-Based Co-operaTive Architecture). *Sensors, 20*(19), 5495.

CHAPTER 8

Design and Optimization of 2-Port MIMO Antennas with Reduced Mutual Coupling Using a Genetic Algorithm

SAIYED TAZEN ALI,[1,*] ZAKIR ALI,[1] V. K. SHARMA,[2] and V. K. SINGH[3]

[1]*Department of Electronics and Telecommunication Engineering, IET, Bundelkhand University, Jhansi, Uttar Pradesh, India*

[2]*Department of Electronics and Telecommunication Engineering, Bhagwant University, Ajmer, Rajasthan, India*

[3]*Department of Electronics and Telecommunication Engineering, CSE Ambabai, Jhansi, Uttar Pradesh, India*

*Corresponding author

ABSTRACT

In this modern era, we have a greater number of antennas in limited space, which leads to mutual coupling. This degrades the channel capacity as well. The improvement in channel capacity and mutual coupling is achieved by adding a resonator of length 38 mm and width 2.5 mm between two patches, each with dimensions of 17.6 × 23 mm², which are placed on an FR4 substrate with dimensions of 70 × 70 × 1.58 mm³, situated on a 0.035 mm thick ground plane of the same dimensions. The simulated S_{11} value is found to be below –15 dB, and S_{21} is around –30 dB at a resonance frequency of 3.94 GHz. To optimize the design of the antenna system, we use a Genetic Algorithm. It is a method for searching a global optimization technique. Due to this algorithm, S_{21} is optimized to –60 dB at the same frequency.

Wireless Technology: Advances and Innovations for Smart IoT Applications. Rajesh Kumar Dwivedi, Vinod Kumar Singh, and Raghav Dwivedi (Eds.)

DOI: 10.1201/9781998511495-8

8.1 INTRODUCTION

Due to its selectivity for operating frequency, pattern, and polarization [1–5], Microstrip Patch Antennas (MSAs) in wireless communication systems have acquired remarkable improvements in technology and design approaches in recent decades. Low mutual coupling [6–10] is easily achieved at the base station, where element separations are always several wavelengths apart [11]. As volume restrictions exist, attaining low mutual coupling will be challenging for mobile terminals. Several antenna designs have been proposed to lessen mutual coupling for mobile terminals. Coupling between coplanar and patch antennas may be caused by near fields and surface waves [12]. When one antenna is positioned in another antenna's near field zone, near field coupling occurs. When the antennas are mounted on dielectric substrates with low permittivity, the coupling is high. In certain circumstances, the coupling may lead to deterioration in the antenna's radiating properties. In extremely thin grounded dielectric substrates, it seems that surface waves are very weakly stimulated, whereas space waves predominate and cause substantial coupling when antennas are adjacent to one another. To enhance antenna radiation characteristics, many research projects have been dedicated to overcoming mutual coupling between coupled antennas [13–15]. There are numerous ways to lessen mutual coupling in MIMO antennas [16], but in reality, it is necessary for the performance of microstrip antenna systems because mutual coupling among closely spaced radiating elements affects induced currents, input impedance, and radiation patterns. This reduces the capacity of MIMO systems. Reducing correlation is made easier by physically separating radiating components by a distance longer than half a wavelength. However, because of size restrictions, this technique is impractical for small wireless devices. Figure 8.1 displays the block diagram of MIMO antennas. A worldwide search of the cost surface is made possible by the genetic algorithm (GA), an optimization technique, using the mechanism of random events statistically described by Darwinian evolution [17]. Figure 8.2 depicts the flow chart for the detailed GA procedure.

The survival of the fittest principle is followed by genetic algorithms. It is an algorithm for global optimization that is based on the concepts of natural selection and genetics [18]. According to Charles Darwin, genes are often the determining factor in optimization; a bit string of these genes produces a chromosome, and generation optimization refers to iterative optimization. The parents are two individuals who may produce a new

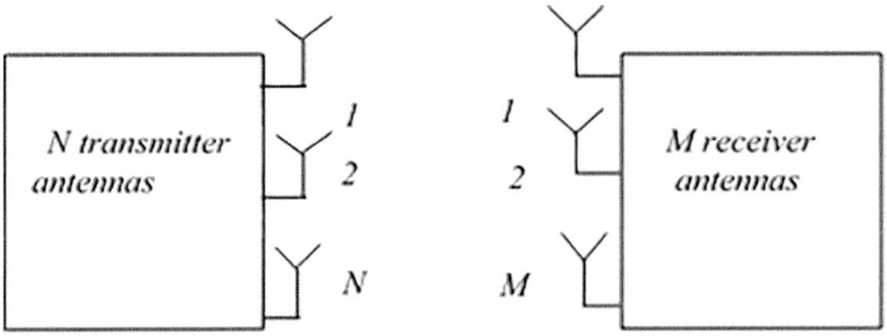

FIGURE 8.1 Block diagram of MIMO antennas.

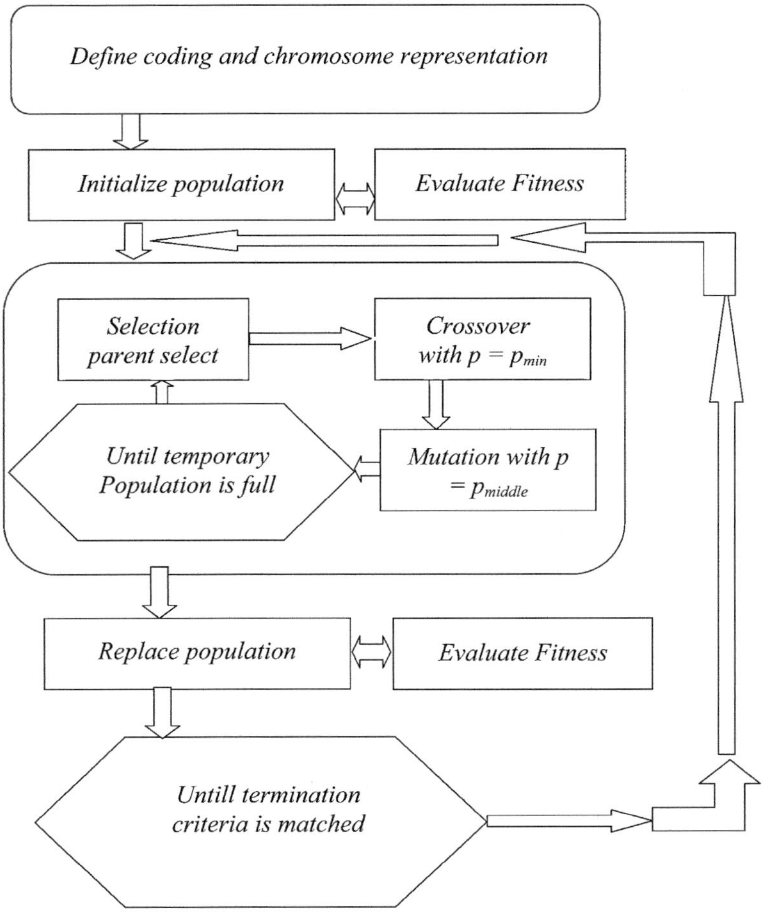

FIGURE 8.2 Flow chart of complete GA process.

individual of the next generation. Then, using the crossover and mutation process, the chosen pair of parents produces offspring. The fitness function is the objective function that defines the optimization objective [19]. The Genetic Algorithm has gained popularity in recent years for its use in determining antenna system optimization values [20]. Genetic Algorithms may also be used to optimize nonlinear multivariate functions.

The genetic algorithm may be used to optimize the length, width, spacing between two patch antennas, and feed positions for each patch antenna [21]. The rectangular patch optimization in GA uses the fitness function from.

8.2 DESIGN OF MIMO ANTENNAS

The rectangular patch antenna of dimension 17.6×23×0.035 mm^3 which is placed on a FR4 substrate whose dimension is 70×70×1.6 mm^3 on a ground plane of same length and width and whose thickness is 0.035 mm. The two rectangular patch antennas separate at a distance of $\lambda_o/2$ from the center of two patches and a resonator of dimensions 38×4×0.035 mm^3. The antenna is fed by insight line feed of dimension 38×4 mm^2. The whole system is analyzed through a CST simulator.

8.2.1 DESIGN PROCEDURE

In this chapter, the focus is on the design and simulation of a two-port Multiple Input Multiple Output (MIMO) antenna system using an FR4 substrate with a permittivity of 4.4. The antenna dimensions are specified as 70×70×1.58 mm^3, and it is situated on a 0.035 mm thick ground plane with the same dimensions, as depicted in Figures 8.3(a) and 8.3(b). The antenna structure involves a single patch unit with dimensions of 17.6×23×0.035 mm^3 placed on the substrate. Subsequently, an identical patch configuration is positioned adjacent to the single patch, and a resonator with dimensions of 38×4 mm^2 is introduced, as illustrated in Figure 8.3(c). The entire system is subjected to simulation in CST Studio, both with and without the resonator, at a resonance frequency of 3.94 GHz. The line feed utilized for the two-port MIMO antenna system is 31.4×4 mm^2. Table 8.1 provides a comprehensive overview of the design parameters for the MIMO antennas, offering a detailed reference for understanding and analyzing the antenna system's performance. This investigation aims to contribute valuable insights into the optimization and functionality of

MIMO antennas for wireless communication applications. To evaluate the performance of the proposed MIMO antenna system, comprehensive simulations are conducted using CST Studio. The simulations compare the system's behavior with and without the resonator at a resonance frequency of 3.94 GHz, allowing for a thorough understanding of the impact of the resonator on the antenna's resonance and radiation properties.

For efficient signal transmission and reception, a line feed of 31.4×4 mm² is employed to feed the two-port MIMO antenna system. This strategic choice of feed line dimensions is a critical aspect of optimizing the antenna's impedance matching and ensuring robust communication capabilities. Table 8.1 presents a detailed compilation of the design parameters for the MIMO antennas, offering a comprehensive reference guide for researchers and engineers interested in the intricacies of the proposed antenna system. This research aims to contribute valuable insights into the realm of MIMO antenna design, providing a foundation for further advancements in wireless communication technologies. The results of this study hold significance in enhancing the performance and adaptability of MIMO antennas for diverse communication applications.

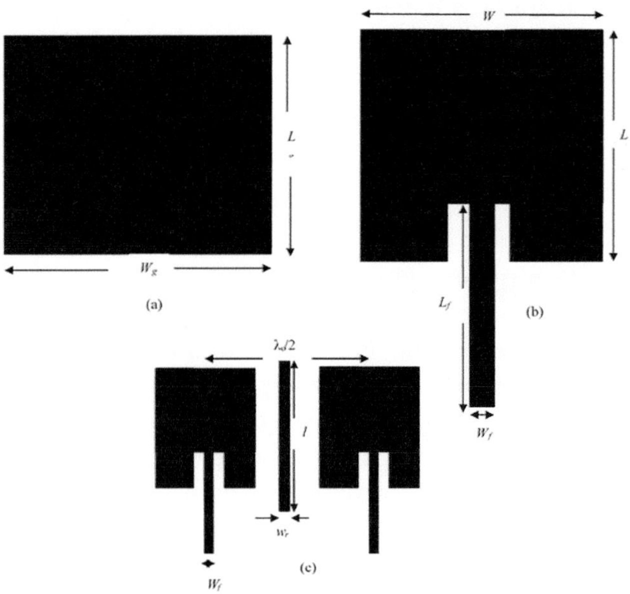

FIGURE 8.3 Two port MIMO antenna with resonator insert at the middle. (a) Defected ground plane (Back view) (b) Unit antenna element with line feed (c) Top view of MIMO antenna system with resonator.

TABLE 8.1 MIMO Antennas Parameters

Parameters	Unit in mm
L_g	70
W_g	70
L	17.6
W	23
L_f	31.4
W_f	4
g_p	38
l_r	38
w_r	4

8.3 RESULTS AND DISCUSSION

The simulated value of S_{11} is found to be –12 dB, –25 dB, and –23 dB for a single unit patch, 2-port MIMO antennas without a resonator, and after adding the resonator, respectively, at 3.94 GHz, as shown in Figure 8.4. The S_{21} values are found to be less than –50 dB without a resonator, and after adding the resonator, it is found to be –60 dB at the 3.94 GHz resonance frequency, which is shown in Figure 8.5. The gain in the H-plane is found to be 3.75 dB before and after adding a resonator, as shown in Figure 8.6. The ECC value calculated from S parameters through post-processing in CST is also shown in Figure 8.7 and is found to be less than 0.01 at the 3.94 GHz resonance frequency.

8.4 OPTIMIZED USING GA

The survival of the fittest principle is followed by genetic algorithms. It is an algorithm for global optimization that is based on the concepts of natural selection and genetics [18]. According to Charles Darwin, genes are often the determining factor in optimization; a bit string of these genes produces a chromosome, and generation optimization refers to iterative optimization. The parents are two individuals who may produce a new individual of the next generation. Then, using the crossover and mutation processes, the chosen pair of parents produces offspring. The fitness function is the objective function that defines the optimization objective [19]. The Genetic Algorithm has gained popularity in recent years for its use in

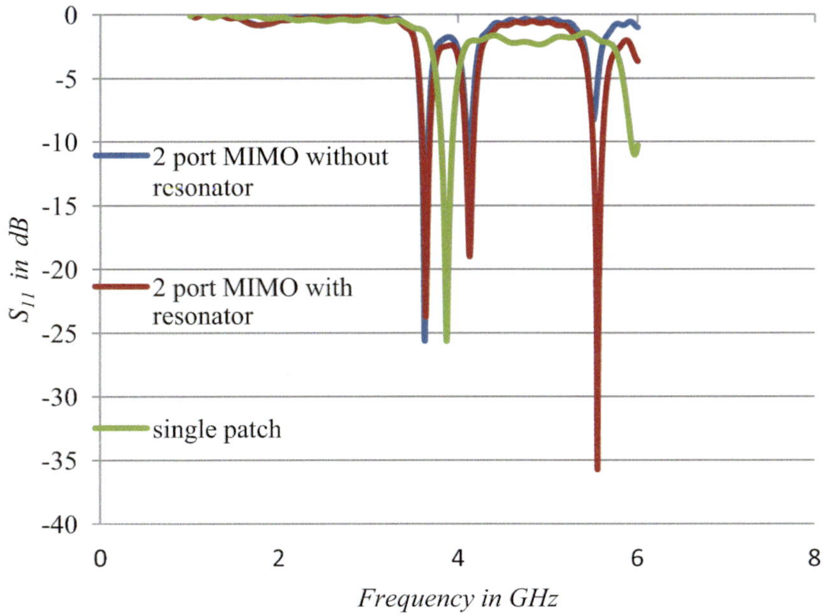

FIGURE 8.4 The S_{11} parameter of two port MIMO antenna without and with resonator.

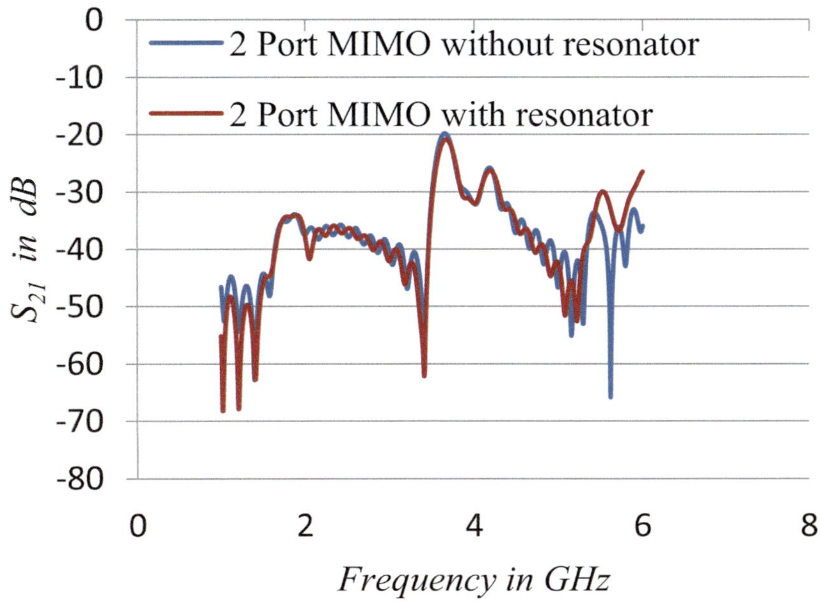

FIGURE 8.5 The S_{21} parameter of two port MIMO antenna without and with resonator.

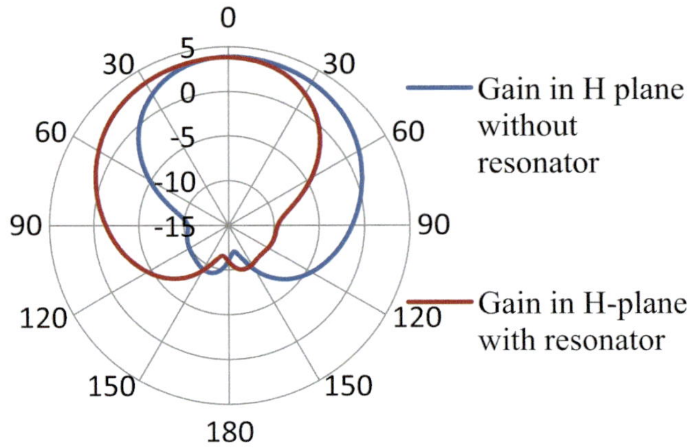

FIGURE 8.6 The gain of two port MIMO antenna without and with resonator in E plane.

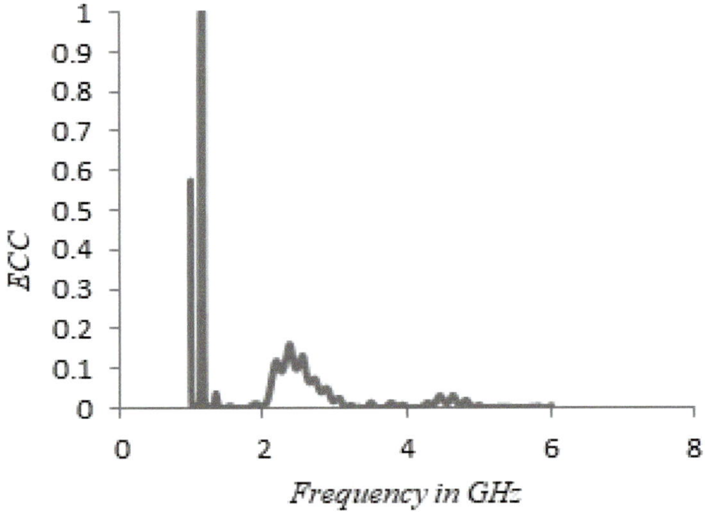

FIGURE 8.7 The ECC calculated from S parameters of two port MIMO antennas.

determining antenna system optimization values [20]. GAs may also be used to optimize nonlinear multivariate functions.

The genetic algorithm may be used to optimize the length, width, spacing between two patch antennas, and feed positions for each patch antenna [21]. The rectangular patch optimization in GA uses the fitness function.

$$f_0 = \frac{c}{2(L+\Delta W)\sqrt{\varepsilon_{eff}(W)}} \tag{8.1}$$

Where c is the speed of light, $\in_r$ is the relative dielectric permittivity, W is the width of patch and f_0 is the resonant frequency.

$$f_0 = \frac{c}{2W\sqrt{\frac{\in_r + 1}{2}}} \tag{8.2}$$

$$\in_{eff}(W) = \frac{\in_r + 1}{2} + \frac{\in_r - 1}{2\sqrt{1 + \frac{10h}{W}}} \tag{8.3}$$

$$\Delta w = 0.412h \frac{[\in_{eff}(w) + 0.300](\frac{w}{h} + 0.264)}{[\in_{eff}(w) - 0.258](\frac{w}{h} + 0.813)} \tag{8.4}$$

$$g_p = \frac{c}{2f_0} \tag{8.5}$$

Where f_0 $\sqrt{\in_{eff}(W)}$ is the effective permittivity, h is the height of the substrate, g_p is the gap between patches, and L is the length of the patch. The Equations 8.1, 8.2, and 8.5 are taken as a fitness function and solved for optimization through a Multi-objective Genetic Algorithm. The optimized results are shown in Table 8.2. The optimized length of the patch, width of the patch, gap between two patches, and position of the feed from the center of the patch are found to be 17.17 mm, 23 mm, 48.50 mm, and 4.6 mm, respectively. The number of iterations, population size, crossover fraction, and probability of mutation are 112, 23, 0.8, and 0.01, respectively. The detailed process of the Genetic Algorithm is shown in Figure 8.2.

TABLE 8.2 Optimization Parameters

Sl. No.	**Input Variable** f_o	**Output Variables** g_p	W	L
1	4.19	47.21	22.00	17.00
2	4.18	47.21	22.03	17.00
3	3.97	48.65	23.00	17.20

TABLE 8.2 *(Continued)*

Sl. No.	Input Variable		Output Variables	
	f_o	g_p	W	L
4	3.98	49.33	22.97	17.50
5	4.06	49.87	22.51	17.70
6	4.02	49.70	22.72	17.62
7	4.05	50.16	22.58	17.85
8	4.04	48.19	22.61	17.06
9	4.13	47.97	22.28	17.21
10	4.10	48.78	22.30	17.26
11	4.09	48.68	22.33	17.23
12	4.14	47.82	22.22	17.15
13	4.16	48.25	22.00	17.16
14	4.11	49.03	22.25	17.43
15	4.03	50.31	22.64	17.82
16	4.07	50.32	22.42	17.83
17	4.19	47.21	22.00	17.00
18	4.02	50.07	22.70	17.73
19	4.08	48.57	22.36	17.18
20	3.99	48.61	22.87	17.20
21	4.01	49.30	22.80	17.48
22	4.01	50.81	22.75	18.00
23	3.97	48.50	23.00	17.17

8.5 CONCLUSION

Table 8.1 provides a comprehensive insight into the optimization process for the two-port MIMO antenna system, where critical parameters like patch length, patch width, feed position, and the gap between antennas are successfully optimized using the GA. GA proves to be an effective optimization tool in navigating the complex design space and identifying optimal values for these parameters, ensuring the antenna system operates at its peak performance. Furthermore, GA extends its utility beyond parameter optimization to enhance the overall compactness of the system. The incorporation of a Multi-objective Genetic Algorithm aids in streamlining the design, striking a balance between conflicting objectives and delivering a solution that is not only optimized but also space-efficient.

This underscores the versatility of GA in addressing multiple design criteria simultaneously, a crucial aspect in modern antenna design where space constraints and performance requirements often pose challenging trade-offs. In the context of this research, the performance of the two-port MIMO antenna system is meticulously examined both before and after the introduction of a resonator between antennas. The resulting S_{11} and S_{21} parameters, indicative of reflection and transmission characteristics, exhibit substantial improvement, with values reaching approximately –28 dB and –35 dB, respectively. These outcomes signify enhanced resonance and isolation properties, essential for minimizing signal interference and maximizing communication reliability. Following the observation of the antenna's performance, the Genetic Algorithm is deployed to further refine the system. Through iterative optimization, GA fine-tunes the design parameters to achieve an optimized configuration. This optimized result reflects the success of GA in tailoring the antenna system to meet specific performance objectives, showcasing its efficacy in the intricate task of antenna design optimization.

In conclusion, this chapter not only demonstrates the efficiency of the Genetic Algorithm in optimizing individual parameters but also highlights its role in addressing the broader challenge of achieving a well-balanced and compact two-port MIMO antenna system. The successful application of GA underscores its significance in advancing the field of antenna design and optimizing the performance of wireless communication systems.

KEYWORDS

- **FR4 substrate**
- **genetic algorithm**
- **global optimization**
- **MIMO antennas**
- **mutual coupling**
- **patch antennas**
- **resonator**

REFERENCES

1. Hammerstad, E. O. (1975). Equations for microstrip circuit design. In *1975 5th European Microwave Conference* (pp. 268–272). IEEE.
2. Bahl, I. J., & Bhartia, P. (1980). *Microstrip Antennas*. Artech House.
3. Fong, K. S., Withers, M. J., & Pues, H. F. (1985). Wideband multilayer coaxial-fed microstrip antenna element. *Electronics Letters, 21*, 497–499.
4. Balanis, C. A. (2012). *Advanced Engineering Electromagnetics*. John Wiley & Sons.
5. Garg, R. (2001). *Microstrip Antenna Design Handbook*. Artech House.
6. Girish, K., & Ray, K. P. (2003). *Broadband Microstrip Antennas*. Artech House Antennas and Propagation Library.
7. Balanis, C. A. (2016). *Antenna Theory: Analysis and Design*. John Wiley & Sons.
8. Antennas, M. (1995). *Microstrip Antennas: The Analysis and Design of Microstrip Antennas and Arrays*.
9. Targonski, S. D., Waterhouse, R. B., & Pozar, D. M. (1996). Wideband aperture coupled stacked patch antenna using thick substrates. *Electronics Letters, 32*(21), 1941–1942.
10. Luk, K. M., Lee, K. F., & Chow, Y. L. (1998). Proximity-coupled stacked circular-disc microstrip antenna with slots. *Electronics Letters, 34*(5), 419–420.
11. Jensen, M. A., & Wallace, J. W. (2004). A review of antennas and propagation for MIMO wireless communications. *IEEE Transactions on Antennas and Propagation, 52*(11), 2810–2824.
12. Mbonjo, H. N. M., Hansen, J., & Hansen, V. (2004). MIMO capacity and antenna array design. In *IEEE Global Telecommunications Conference, 2004. GLOBECOM'04* (Vol. 5, pp. 3155–3159). IEEE.
13. Jiang, H., Si, L. M., Cheng, G., & Lv, X. (2019). A wideband close-spaced 5G MIMO antenna with mutual coupling reduction using combined decoupling networks. In *2019 International Conference on Microwave and Millimeter Wave Technology (ICMMT)* (pp. 1–3). IEEE.
14. Kaiser, T., Zheng, F., & Dimitrov, E. (2009). An overview of ultra-wide-band systems with MIMO. *Proceedings of the IEEE, 97*(2), 285–312.
15. Svantesson, T., & Ranheim, A. (2001). Mutual coupling effects on the capacity of multielement antenna systems. In *2001 IEEE International Conference on Acoustics, Speech, and Signal Processing. Proceedings* (Cat. No. 01CH37221) (Vol. 4, pp. 2485–2488). IEEE.
16. Paulraj, A. J., Gore, D. A., Nabar, R. U., & Bolcskei, H. (2004). An overview of MIMO communications-a key to gigabit wireless. *Proceedings of the IEEE, 92*(2), 198–218.
17. Rahmat-Samii, Y. (2007). Modern antenna designs using nature inspired optimization techniques: Let Darwin and the bees help designing your multi band MIMO antennas. In *2007 IEEE Radio and Wireless Symposium* (pp. 463–466). IEEE.
18. Uthansakul, P., Assanuk, D., & Uthansakul, M. (2010). The use of Genetic Algorithm for designing MIMO antenna placement. In *2010 International Conference on Computer and Information Application* (pp. 414–417). IEEE.

19. Song, S., & Murch, R. D. (2013). An efficient approach for optimizing frequency reconfigurable pixel antennas using genetic algorithms. *IEEE Transactions on Antennas and Propagation, 62*(2), 609–620.
20. Weile, D. S., & Michielssen, E. (1997). Genetic algorithm optimization applied to electromagnetics: A review. *IEEE Transactions on Antennas and Propagation, 45*(3), 343–353.
21. Pattnaik, S. S., Khuntia, B., Panda, D. C., Neog, D. K., & Devi, S. (2003). Calculation of optimized parameters of rectangular microstrip patch antenna using genetic algorithm. *Microwave and Optical Technology Letters, 37*(6), 431–433.

CHAPTER 9

Analysis and Study of 5G mmWave MIMO Antenna Array

MANISH SHARMA[1,*] and MANOJ KAPIL[2]

[1]*Chitkara University Institute of Engineering and Technology, Chitkara University, Punjab, India*

[2]*Department of Computer Science, Houston Tillotson University, Texas, USA*

*Corresponding author

ABSTRACT

This chapter focuses on the study of a 5G millimeter wave antenna designed on a low permittivity substrate to achieve high impedance matching. The 5G communication requires complicated infrastructure for its implementation. The congestion at the lower band in the microwave energy band has prompted the FCC to allocate a millimeter bandwidth of 25.0 GHz–70.0 GHz for 5G mmWave. Numerous radiating patch antennas within the mmWave band have been reported with radiating shapes such as square, slotted circular, orthogonal slotted patch, Y-shaped slotted patch, and so forth. These antennas are transformed into array configurations to achieve high gain and are also converted to MIMO configurations intended for information and communication technology to achieve spatial diversity. The MIMO technology antenna array achieves high isolation by placing the radiating elements in an orthogonal sequence, at an angle of 180°, by using etched rectangular strips in the ground, or by using novel decoupling parasitic elements. Future mobile users will experience much faster communication with the development of 5G technology.

Wireless Technology: Advances and Innovations for Smart IoT Applications. Rajesh Kumar Dwivedi, Vinod Kumar Singh, and Raghav Dwivedi (Eds.)

DOI: 10.1201/9781998511495-9

9.1 INTRODUCTION

The breakthrough in modern wireless communication has led to the advancement of antenna systems from single-input-single-output (SISO) to multiple-input-multiple-output (MIMO). The single antenna system, both at the transmitter and receiver, does suffer drawbacks such as multipath fading effects in real-time wireless communication channels. To overcome this drawback, multiple antennas are used at both the transmitter and receiver sides. The Federal Communications Commission (FCC) allocates different working bands to target the need from time to time. One such allocation has been proposed by the FCC for millimeter-wave applications with bandwidth ranging between 25 GHz and 70 GHz [1]. This bandwidth allocation corresponds to the 5th generation (5G), which offers greater flexibility for antenna designers and features such as miniaturization, achieving good far-field results, and high diversity performance. To achieve this, the MIMO antenna system needs to be developed for faster communication. The objective can be achieved at millimeter wave bands by achieving multi-resonance or wide bandwidth from millimeter wave FCC bands. A hexagonal patch antenna etched with a rectangular slot and a rectangular ground printed on the opposite plane of Rogers 4003C substrate helps in achieving a wider bandwidth of 26 GHz–38 GHz with two useful millimeter wave bands: 26 GHz–29 GHz and 36 GHz–41 GHz [1]. The antenna is further converted to a 2×2 MIMO configuration. A famous-element MIMO antenna system fabricated on Rogers RT Duroid 5880 substrates with a height of 0.762 mm features a unique radiating patch with a five-sided polygon forming patch and an identical shape etched on the ground. In addition, two triangular stales are added as parasitic elements, which overall achieve resonances at 23.50 GHz and 27.8 GHz, respectively [2]. The 5G evaluation releases define the release of 5G NR-16, which features a MIMO system with the ability of beamforming, sharing of dynamic spectrum with dual connection, and aggregation of the carriers [3]. A low-frequency application 2×2 MIMO antenna achieves polarization diversity with dual bands: 2.408 GHz–2.776 GHz and 4.960 GHz–5.640 GHz. All four elements are placed in an orthogonal configuration, which is the reason for achieving polarization diversity [4]. A 28 GHz MIMO antenna with a compact size of 11 mm × 31 mm operates in the bandwidth of 26 GHz – 31 GHz, providing isolation of more than 20 dB [5]. The evolution of 5G technology, which utilizes orthogonal frequency division multiple access (OFDMA), starts from the 1st generation with frequency modulation, followed by the 2nd generation with applications in GSM. The third generation involves the universal mobile telecommunications system (UMTS) and long-term evolution (LTE) applications in 4G bands [6]. The 4G LTE system offers a downlink speed of 100 Mbps and an uplink speed of 50 Mbps. The MIMO system offers several advantages, such as optimized values of Bit Error Rate (BER) and reduced fading [7]. A dual-band MIMO antenna resonating at 28 GHz and 38 GHz is useful for 5G high-speed communication, where the radiating element is placed in a mirrored and adjacent fashion [8]. An E-shaped 5G mm-wave dual-band antenna offers circular polarization at 28 GHz and 38 GHz. The design also utilizes L-shaped and hat-shaped metallic strips fabricated on Rogers RO 5880 microwave substrate [9]. A millimeter wave circular slotted antenna resonates at 28 GHz with an operating bandwidth of 26.16 GHz–29.72 GHz [10]. A multi-circular loop antenna array is designed on Rogers RO5880 substrate with a thickness of 0.254 mm, offering a wide impedance bandwidth of 26.50 GHz–40 GHz. The size of the array is very compact, with dimensions of 18.25 mm × 12.5 mm, offering a peak gain of 11 dBi [11, 12]. A fork-shaped radiating patch with a slotted rectangular ground offers dual-band resonance with isolation of more than 35 dB [13, 14].

An eight-port MIMO antenna covers the N261 5G bandwidth of 27.50 GHz and offers circular polarization in the same bandwidth [15]. A four-port circular slotted antenna with a rectangular ground provides resonances at 26.843 GHz [16], and a very wide impedance bandwidth of 25.0 GHz–39.0 GHz is achieved on ultra-thin Rogers RO5880 substrate with a compact dimension of 24 × 24 mm^2, producing a gain of 7.10 dBi [16, 17, 19]. A single-band MIMO antenna resonating at 28 GHz converted to a MIMO array achieves a gain of 12.0 dBi [18–20]. A 37 GHz MIMO antenna has a compact size of 6.84 dB [21, 22]. A MIMO antenna of compact size 27.65 × 12 mm^2 achieves dual bandwidths of 26 GHz–30 GHz and 36 GHz–41.5 GHz, with isolation of more than 20 dB, corresponding to gains of 5.2 dBi and 5.3 dBi [23]. Several sectors, such as financial sectors, telecommunications, automotive, and so forth, are a few areas that are addressed by the 5G communication channel [24].

9.2 THE 5G MMWAVE ANTENNA

The communication link between the transmitter and the receiver is achieved by using an antenna system. Also, the mmWave compact antennas have emerged as an option to realize microwave frequency bandwidth. In today's scenario, features such as low latency with high data rates have led to healthy competition between service providers. It is expected that in the near future, the speed will be doubled, and approximately 70% of the global population will use the emerging 5G technology in the millimeter wave bandwidth. Additionally, the quality of service is expected to increase through the use of MIMO system implementation. The overall performance of the communication system is enhanced by implementing the MIMO system due to reasons such as the SISO system suffering from fading, which thereby reduces the operating bandwidth and efficiency.

The MIMO planar antenna at mmWave has taken a lead role in communication due to the advanced technology development in the area of microwave-integrated-circuit (MIC) and monolithic-MIC (MMICs). The SISO system does suffer from fading in the communication channel, which results in the degradation of the received signal, which is quantified by the signal-to-noise (SNR) ratio and also includes a higher probability of error. This problem is tackled by using diversity schemes such as temporal diversity, frequency diversity, and antenna diversity.

The above three diversity schemes are achieved by using SISO, MISO, and MIMO configurations. In the temporal diversity scheme, the transmitted signal utilizes different instances of time and uses error correction with bit interleaving (to correct errors). In the frequency diversity scheme, frequency hopping and Orthogonal-Frequency-Division-Multiplexing (OFDM) techniques are used. The popular diversity scheme, antenna diversity, offers diversity schemes such as spatial diversity, radiation pattern diversity, and polarization diversity. The mmWave antenna discussed utilizes one of the antenna diversity schemes, which are discussed below.

Figure 9.1 shows the configuration of the multiple-input-multiple-output 2×2 configuration. The antenna shown in Figure 9.1 is designed with reference number [1] for 5G-mmWave applications covering two bands centered at 27 GHz and 39 GHz, respectively. The antenna configuration is iterated in three steps, with step 1 including a rectangular patch of dimensions 3.65 mm×2.92 mm and a rectangular ground printed on Rogers RO4003C microwave substrate with permittivity $\varepsilon_r = 3.55$. The first iteration results in a single resonance centered at 30 GHz with $S_{11} = -15.0$ dB. The second iteration is the modified version with an additional two triangles added at the edges of the rectangular patch with an edge dimension of 2.92 mm. This modification further improves the impedance matching at 30 GHz with $S_{11} = -25.12$ dB. The dual-band with resonance frequencies at 27.0 GHz and 37.0 GHz is achieved by etching a rectangular slot of dimensions 3.80 mm×2.34 mm, as shown in Figure 9.1(a)-(b). The optimized dimension of the ground is 6.60 mm, and the MIMO configuration of the above dual band is achieved by adding an identical antenna rotated by 180° as shown in Figure 9.1(c)-(d). The minimum distance of 5.00 mm is maintained between both radiating antennas for better isolation. The MIMO antenna observes isolation of more than 25 dB and inherits characteristics such as a gain of 5.00 dBi and 5.70 dBi at 27.0 GHz and 39.0 GHz, respectively. It is also to be noted that the MIMO antenna offers efficiencies of 99.5% at 27.0 GHz and 98.6% at 39.0 GHz, respectively.

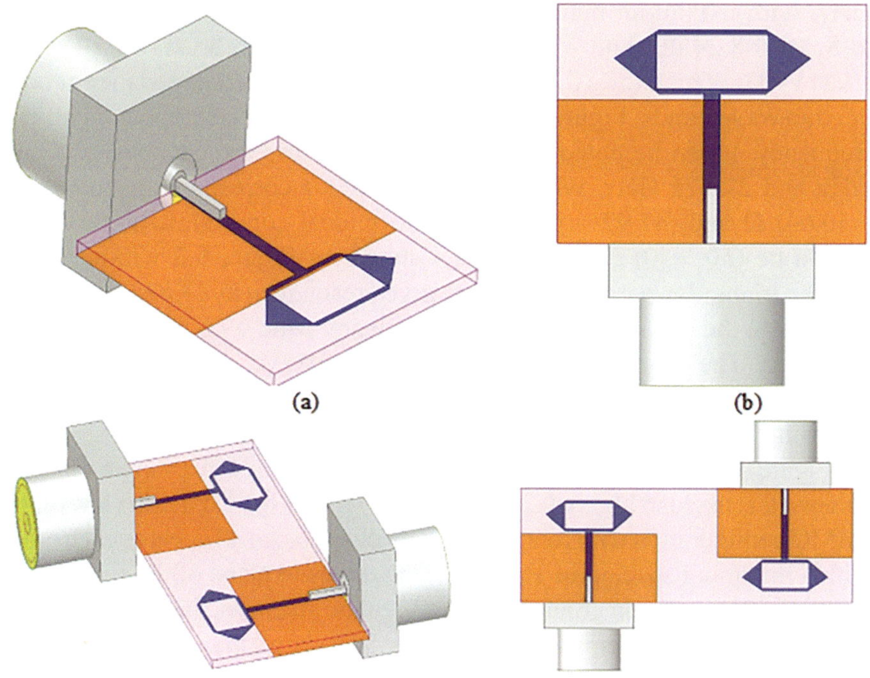

FIGURE 9.1 Dual-band antenna: (a) Isometric view, (b) Front view (radiating patch and ground), (c) Isometric view of two-port MIMO antenna, (d) Front view of two-port MIMO antenna.

Figure 9.2 shows the development of a single-element antenna for the wide frequency range of 23.0 GHz–29.0 GHz applications [2]. Figure 9.2(a)-(c) shows the isometric, radiating, and rear-ground views. The radiating patch is printed on the top plane of the Rogers RTDuroid 5880 microwave substrate with a permittivity of 2.20 and grounded on the opposite plane with a Defected Ground Structure (DGS). The dimensions of the antenna are very compact, with an overall volume of 15 × 19.9 × 0.762 mm^3. Figure 9.2(a) also shows a 50 Ω SMA connector that is soldered to a matched 50 Ω feedline for input signal transmission. Figure 9.2(b) reflects the radiating patch with a Y-shaped design, and two triangular parasitic patches are also placed. The ground plane printed on the opposite side is defective by etching a similar Y-shape to that of the radiating patch. Figure 9.2(d) shows the S_{11} simulated graph of the 5G mmWave antenna. The antenna offers dual-band operation with a bandwidth of 22.31 GHz–24.40 GHz, with maximum matched impedance at 23.42 GHz, corresponding to S_{11} = –48.24 dB, and 25.63 GHz–28.31 GHz, with S_{11} = –32.14 dB at 27.215 GHz. This shows that the antenna offers matched impedance at both 5G mmWave bands. Figure 9.2(e) shows the impedance graph concerning frequency. It can be noted that the two resonances are centered at 23.42 GHz and 27.215 GHz, with impedance values corresponding to (49.62 + j0.04) Ω and (47.97 – j1.31) Ω. Under ideal conditions, these values should be (50 + j0) Ω. The antenna shown in Figure 9.2 without DGS and the triangular-parasitic structure gives resonance at 23.82 GHz, while with DGS, the antenna offers a dual frequency band with poor impedance matching. The objective of the designed mmWave antenna is achieved by adding two rectangular parasitic elements, providing wider impedance matching between 22.76 GHz and 29.13 GHz.

Figure 9.3(a)-(c) [2] shows the MIMO version of the single-element antenna discussed in Figure 9.2. Figure 9.3(a) shows the isometric view of the MIMO configuration with four identical patches placed adjacent to each other with a minimum spacing of 15.0 mm. The opposite plane of the substrate is printed with a common ground, and between the antennas, three slots are also etched to achieve higher isolation. Figure 9.3(c) shows the ground plane with three etched slots. The antenna provides a wider impedance bandwidth of 22.294 GHz–28.320 GHz, noted in Figure 9.3(d). The antenna also resonates with dual frequencies centered at 23.50 GHz and 27.40 GHz, respectively. Figure 9.3(e) shows the isolation between the radiating elements, where the first element is excited, and the coupling effects are observed by studying the

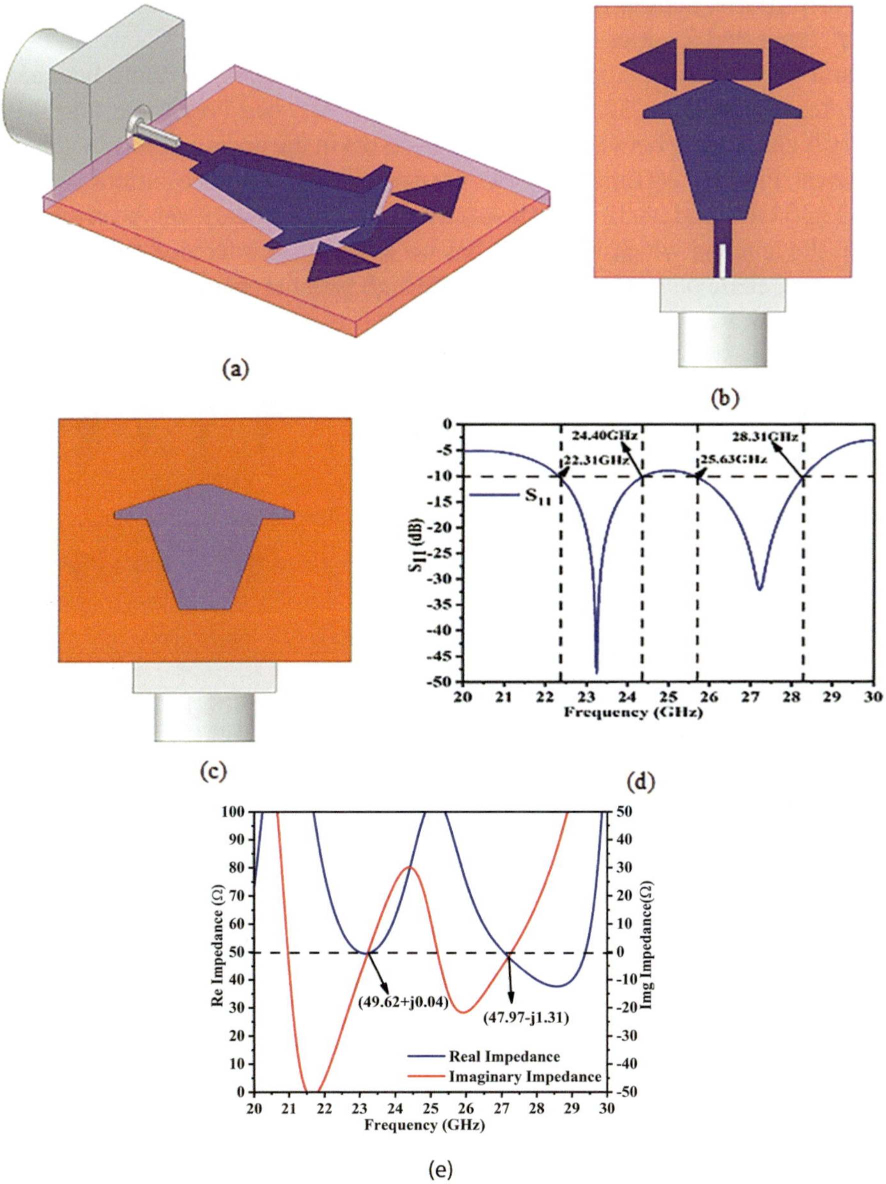

FIGURE 9.2 Broadband mmWave antenna: (a) Isometric view, (b) Front view, (c) Ground view, (d) S_{11} graph, (e) Impedance graph.

surface current density distribution. It can be seen that the isolation between Port 1 and Port 2 varies between –32 dB and –40 dB, whereas between Port 1 and Port 3, the isolation improves due to the total inter-spacing between element distances of 30.00 mm. The isolation between Port 1 and Port 4 is very high, ranging between –35 dB and –55 dB in the operating bandwidth of interest. Figure 9.3(f) and (g) shows the distribution of surface current density at 22.30 GHz and 28.32 GHz, respectively. In both cases, it can be observed that the antenna which is excited has no effect of interference between the radiating elements, and this is due to the high isolation between the radiating elements shown in Figure 9.3(e).

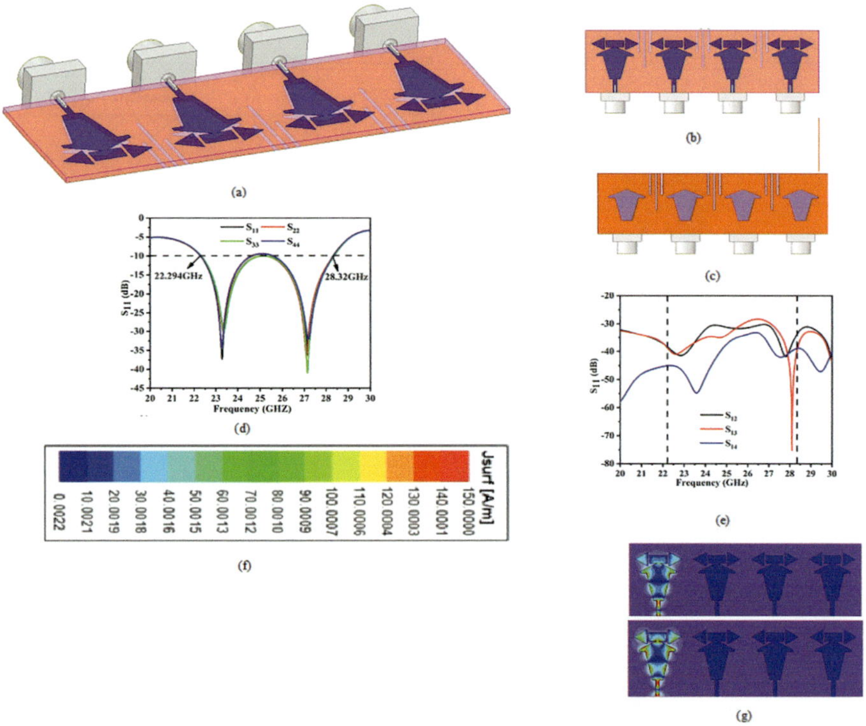

FIGURE 9.3 4-port MIMO broadband mmWave antenna: (a) Isometric view, (b) Front view, (c) Ground view; S-parameter graph: (d) Reflection coefficient, (e) Transmission coefficient; surface current density distribution at (f) 22.30 GHz, (g) 28.32 GHz.

Figure 9.4 shows the single- and four-port MIMO antenna, which offers dual resonating at 2.40 GHz and 5.20 GHz [4]. The radiating patch with rectangular stubs also uses a partial stepped ground. The lower frequency bands are achieved by connecting a series of rectangular stubs that form the letter G, as shown in Figure 9.4(a). The 4×4 combination is obtained by placing the radiating element and respective ground in an orthogonal sequence, which is shown in Figure 9.4(b). In the SISO-MIMO configuration, the two bands are 2.348 GHz–2.600 GHz and 4.712 GHz–4.972 GHz. The MIMO configuration offers isolation of more than 20.0 dB in both operating bands, with the overall size in MIMO configuration being 70×70 mm².

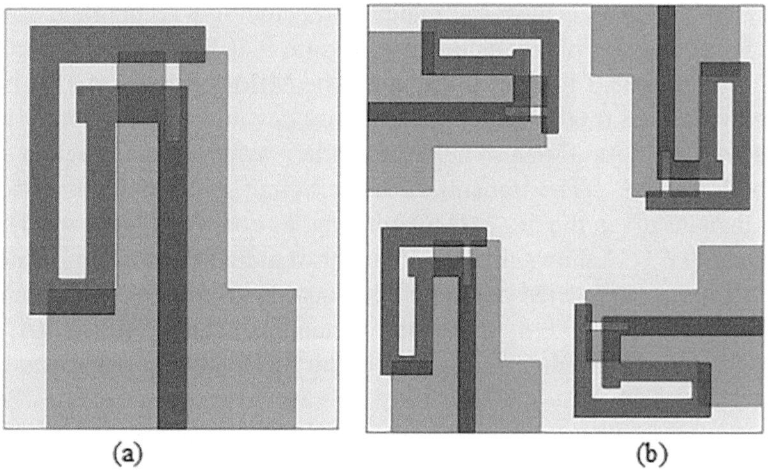

FIGURE 9.4 (a) Single-port low-frequency dual-band antenna, (b) Four-port MIMO configuration.

Figure 9.5 shows the mmWave antenna designed for 28 GHz 5G applications. Figure 9.5(a) shows the initial design with a circular patch and complete ground, which provides resonance at 28 GHz with S_{11} = –16.30 dB. This resonance is governed by the following Equations 9.1–9.3 given by [10].

$$R_{eff} = R\left(\sqrt{1+\frac{2H}{\pi\varepsilon_r R}(ln\frac{\pi R}{2H}+1.7726)}\right) \tag{9.1}$$

$$R = \frac{F}{\left(\sqrt{1+\frac{2H}{\pi\varepsilon_r R}(ln\frac{\pi F}{2H}+1.7726)}\right)} \tag{9.2}$$

$$F = \frac{8.791\times10^9}{f_r\sqrt{\varepsilon_r}} \tag{9.3}$$

where R is the radius of the patch, R_{eff} is the effective radius, H is the height of the substrate, εr is the permittivity of the substrate, and F is a constant value.

Further modification of the 5G antenna is initiated to improve the resonance value corresponding to the matching of impedance. Two rectangular slots are etched on the circular patch, as shown in Figure 9.5(b), which further improves the matching of impedance. This configuration, shown in Figures 9.5(a)–(c), corresponds to a single-element system. Further, the MIMO configuration of the single element is achieved by placing the radiating elements in an orthogonal manner and etching 8 rectangular slots on each partial ground plane, as noted in Figure 9.5(d). Figure 9.5(e) shows the S_{11} graph for all four ports excited. The MIMO antenna maintains all the individual 28.0 GHz bands without any deviation. On the other hand, the MIMO configuration also achieves higher isolation of more than –23.0 dB for S_{12} and S_{13}. The transmission coefficient graph notes isolation of more than 27 dB in the 5G-MIMO antenna, as noted in Figure 9.5(f).

Figure 9.6 [11] shows the development of a dual-band MIMO antenna. Two iterations are carried out to achieve dual-band resonance at mmWave applications. Figure 9.6(a) shows the rectangular patch antenna #A, with width W_{mmW} and length L_{mmW} calculated by the following equations given below [11].

$$Wmmw = \frac{c}{2fr\sqrt{\frac{\varepsilon r+1}{2}}} \tag{9.4}$$

$$\varepsilon_{eff} = \frac{\varepsilon r+1}{2} + \frac{\varepsilon r-1}{2}\left[\frac{1}{\sqrt{1+12\frac{H}{W}}}\right] \tag{9.5}$$

$$Lmmw = \frac{c}{2fr\sqrt{\varepsilon_{eff}}} - 0.824 \times H \frac{\left[(\varepsilon_{eff}+0.30)\left(\frac{W}{H}+0.264\right)\right]}{\left[(\varepsilon_{eff}+0.258)\left(\frac{W}{H}+0.80\right)\right]} \tag{9.6}$$

where C is the speed of light = 3×10^8 m/s, ε_r is the permittivity of the microwave substrate, H is the height of the substrate, W is the width of the transmission line, ε_{reff} is the effective permittivity, and fr is the resonance frequency in GHz. The Antenna A provides a single resonance that is

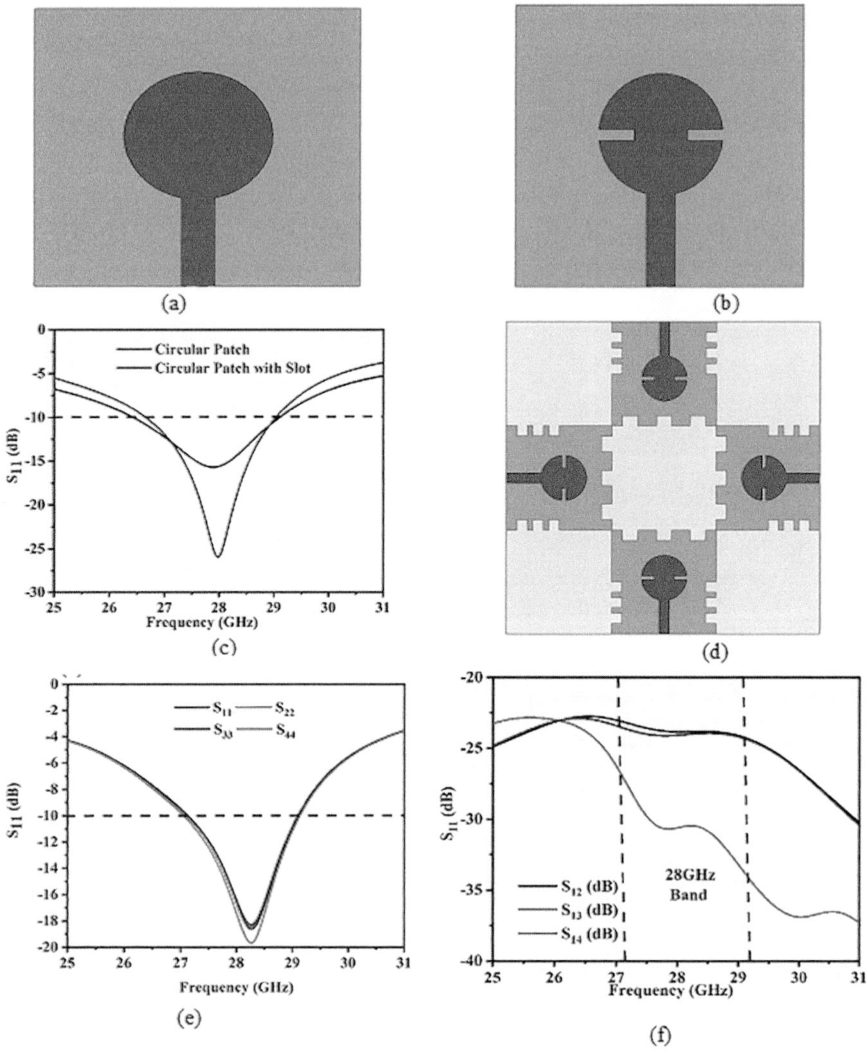

FIGURE 9.5 (a) Single-port circular patch antenna, (b) Slotted circular patch antenna, (c) S-Parameter of antenna #A and antenna #B, (d) Four-port MIMO antenna, (e) Reflection coefficient ($S_{11}/S_{22}/S_{33}/S_{44}$) result, and (f) Transmission coefficient ($S_{12}/S_{13}/S_{14}$).

centered at 30.63 GHz with S_{11} = –10.18 dB. However, the objective is to achieve the mmWave-FCC band of 26.50 GHz–32.90 GHz, and this is achieved by performing two transformations on Antenna A. The first is converting a rectangular patch to a fork-shaped radiator, and the other is etching a rectangular slot on the ground (DGS). This provides the two resonating frequencies centered at 28.0 GHz and 29.80 GHz, respectively, as shown in Figure 9.6(c). The MIMO configuration with corresponding S-parameters is shown in Figure 9.6(d)–(e). The two identical mmWave antennas shown in Figure 9.6(d) form a MIMO configuration with minimal inter-spacing of d = 10.50 mm. The S-parameters ($S_{11}/S_{12}/S_{21}/S_{22}$) shown in Figure 9.6(e) preserve the bandwidth of 26.68 GHz–31.86 GHz (S_{11}/S_{22}) and isolation of more than S_{12}/S_{21} > 32.33 dB.

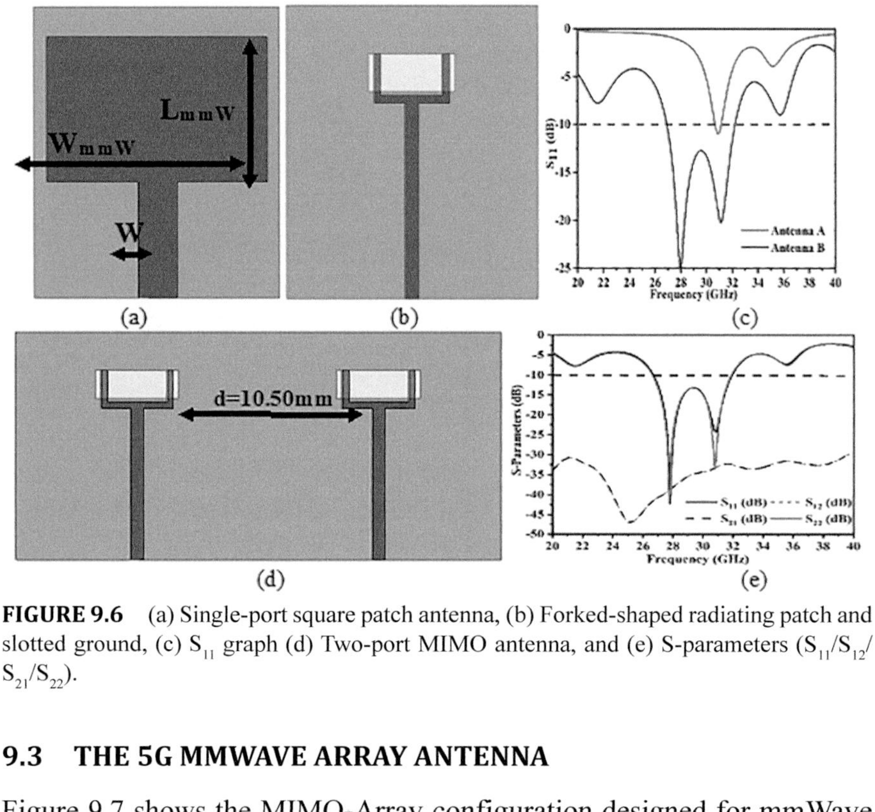

FIGURE 9.6 (a) Single-port square patch antenna, (b) Forked-shaped radiating patch and slotted ground, (c) S_{11} graph (d) Two-port MIMO antenna, and (e) S-parameters ($S_{11}/S_{12}/S_{21}/S_{22}$).

9.3 THE 5G MMWAVE ARRAY ANTENNA

Figure 9.7 shows the MIMO-Array configuration designed for mmWave applications [18, 21]. The two antennas, Antenna #C and Antenna #D, are discussed, which are designed to resonate at 28.0 GHz and 37.0 GHz, respectively. Figure 9.7(a)–(d) shows the single-element Antenna #C resonating at 28.0 GHz. The radiating patch and ground are printed on the opposite plane of the substrate, as shown in Figure 9.7(a). A quarter-wavelength transformer is connected between the transmission line and the radiating patch for matching the impedance. The patch consists of rectangular shapes that are etched with four rectangular slots. The two mirrored triangular slots are etched at the center of the patch, as shown in Figure 9.7(b). The simulated S_{11} graph shown in Figure 9.7(c) resonates at 28.0 GHz with S_{11} = –30.58 dB and covers the bandwidth of 27.33 GHz–28.80 GHz. Figure 9.7(d) shows the impedance graph of the designed antenna. The resonance frequency at 28.0 GHz reads a value of 50 Ω for real impedance and 0 Ω for imaginary impedance. For bandwidth, $\Delta\omega$ = 1.47 GHz, the variation of real and imaginary impedance is very minor. The MIMO-Array version of Antenna #C is shown in Figure 9.7 (i). Firstly, the two elementary elements are formed, which are then placed orthogonally. The isolation is achieved by placing the parasitic structure between the two pairs of the array configuration. The four-port MIMO-array elements preserve the resonance bandwidth and maintain isolation of more than 29.82 dB.

Figure 9.7(e)–(h) shows the Antenna #D configuration. Figure 9.7(e) shows the isometric view, whereas Figure 9.7(f) shows the front view. The designed antenna is printed on the top plane, and the bottom plane consists of the ground. The shape of the patch is like a Y-shape, which is rotated by 90° and is connected to the quarter-wavelength transformer for matching the impedance, which is in turn connected to the 50 Ω transmission line. The single element resonates at 37.02 GHz with S_{11} = –25.50 dB and band impedance between the bandwidth of 36.67 GHz–37.39 GHz. Figure 9.7(h) shows the real and imaginary $\Delta\omega$ = 0.72 GHz. The impedance at 37.02 GHz offers a real value of 50 Ω and an imaginary value of 0 Ω. Figure 9.7(j) shows the array-MIMO configuration of the antenna shown in Figure 9.7(e)–(f). Each MIMO array consists of four radiating patches connected to a microstrip feedline with a fading network. The gap between each array element is 1.90 mm. The width of the transmission line is calculated by [25].

$$W_Z = \left(\frac{377}{Z_c \sqrt{\varepsilon_r}} - 2 \right) \times h_{sub} \tag{9.7}$$

where Z_c is the impedance of the transmission feedline, W_z is the width of the feedline and h_{sub} is the height of the substrate. The reflection coefficient of the 1×4 array offers S_{11}= –33.23 dB at 37.0 GHz with a bandwidth of 677 MHz.

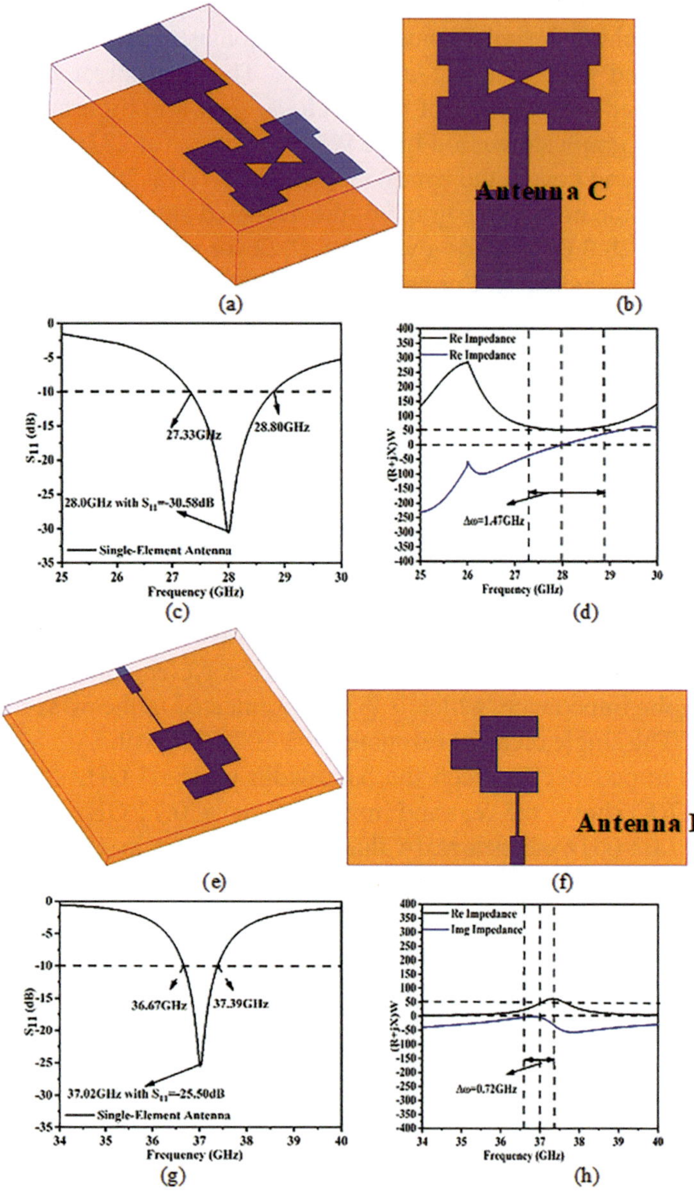

FIGURE 9.7 Single-port square slotted patch antenna for 28.0 GHz-antenna #C: (a) Isometric view, (b) Front view, (c) S_{11} result of antenna, (d) Impedance graph of antenna; patch antenna for 37.0 GHz-Antenna #D, (e) Isometric view, (f) Front view, (g) S_{11} result of antenna, (h) Impedance graph, (i) Array configuration of antenna #C, and (j) Array configuration of antenna #D.

9.4 THE 5G MMWAVE ARRAY-ANTENNA STATE-OF-THE-ART COMPARISON

Table 9.1 shows the comparison of the present state-of-the-art 5G MIMO antenna array. The array MIMO configuration offers high gain when compared to the antenna with MIMO and single element. This is because the radiating signal from the array of elements forms a constructive addition of the electric field intensity. The isolation from Table 7.1 concludes that the absence and presence of decoupling elements offer isolation of more than 20 dB in each case.

TABLE 9.1 Comparison Table of Key Parameters

Sl. No.	Resonance Frequency (GHz) No. of Ports	Size (mm^2) εr	Array Configuration	Gain (dBi)	Isolation (dB)	Diversity Performance (dB)	
						ECC	DG (dB)
1.	27, 39; 2	11×27; 3.55	No	5.00	>30.0	$>1\times10^{-4}$	>9.9995
2.	23.9, 27.3; 4	19.9×60; 2.20	No	7.10	>30.0	NC*	>9.95
3.	28, 38; 1	7.50×8.80; 3.00	No	6.60; 5.86	NA*	NA*	NA*
4.	28, 38; 4	17×20; 2.20	No	12.69; 11.10	NC*	NC*	NC*
5.	28; 4	30×30; 2.20	No	7.10	>30	$>5\times10^{-4}$	>9.999
6.	28.8, 37.9; 1	18.5×22.5; 2.30	Yes	11.50	NA*	NA*	NA*
7.	28; 2	15×30; 3.55	No	5.42	>35.8	$>5\times10^{-4}$	>9.990
8.	28; 4	20×20; 2.20	No	9.24	>18.0	$>13\times10^{4}$	NC*
9.	28, 38; 4	24×24; 2.20	No	7.10	>26.0	$>5\times10^{-2}$	NC*
10.	35; 4	13.75×13.75; 3.55	No	5.00	>30.0	$>3\times10^{-3}$	>9.60
11.	28; 4	15.15×15.50; 2.20	Yes	12.25	>40.0	$>3\times10^{-4}$	>9.90
12.	37; 4	20×40; 2.20	Yes	12.80	>40.0	$>3\times10^{-4}$	>9.95
13.	28, 38; 2	15×25; 2.30	No	8.30	>27.0	$>7\times10^{-3}$	>9.98
14.	28, 38; 4	12×27.65; 3.55	No	5.20; 5.30	>34.0	$>2\times10^{-4}$	>9.99
15.	28; 4	18.5×26.9; 2.20	Yes	10.30	NC*	NC*	NC*

NA*: Not applicable, NC*: Not calculated.

9.5 CONCLUSION

This chapter reports on microstrip transmission line-fed single element, MIMO, and MIMO-array antennas for high-speed wireless communication and high gain. The diversity performance is achieved by placing the radiating elements in an adjacent or orthogonal sequence, which results in spatial diversity. Additionally, the radiating elements are fed by a single source, forming 1×2 or 1×4 configurations, and in general, 1×n (where n=1, 2, 3...n), which results in high gain due to the algebraic addition of the electric field intensity. The impedance is also calculated for each antenna, showing that at resonance, the value of overall impedance is (50+j0) Ω. 5G mmWave antennas were also studied for 28 GHz/38 GHz applications, with characteristics such as S-parameters and impedance matching. A state-of-the-art comparison is also made with the reported work for 5G mmWave applications.

KEYWORDS

- **5G mmWave**
- **5G technology**
- **diversity**
- **microwave substrate**
- **multiple-input-multiple-output**
- **polarization**
- **single-input-single-output**

REFERENCES

1. Ali, W., Das, S., Medkour, H., & Lakrit, S. (2021). Planar dual-band 27/39 GHz millimeter-wave MIMO antenna for 5G applications. *Microsystem Technologies, 27*, 283–292.
2. Sharma, S., Mainuddin, Kanaujia, B. K., & Khandelwal, M. K. (2020). Implementation of four-port MIMO diversity microstrip antenna with suppressed mutual coupling and cross-polarized radiations. *Microsystem Technologies, 26*(3), 993–1000.
3. Peisa, J., Persson, P., Parkvall, S., Dahlman, E., Grøvlen, A., Hoymann, C., & Gerstenberger, D. (2020). 5G evolution: 3GPP releases 16 & 17 overview. *Ericsson Technology Review, 2020*(2), 2–13.
4. Malviya, L., Panigrahi, R. K., & Kartikeyan, M. (2016). A 2× 2 dual-band MIMO antenna with polarization diversity for wireless applications. *Progress In Electromagnetics Research C, 61*, 91–103.

5. Wani, Z., Abegaonkar, M. P., & Koul, S. K. (2018). A 28-GHz antenna for 5G MIMO applications. *Progress In Electromagnetics Research Letters, 78*, 73–79.
6. Alsharif, M. H., & Nordin, R. (2017). Evolution towards fifth generation (5G) wireless networks: Current trends and challenges in the deployment of millimetre wave, massive MIMO, and small cells. *Telecommunication Systems, 64*, 617–637.
7. Farahat, A. E., & Hussein, K. F. (2022). Dual-band (28/38 GHz) wideband MIMO antenna for 5G mobile applications. *IEEE Access, 10*, 32213–32223.
8. Sadeghi-Marasht, S., Sharawi, M. S., & Zhu, A. (2022). Dual-band circularly polarized antenna array for 5G millimeter-wave applications. *IEEE Open Journal of Antennas and Propagation, 3*, 314–323.
9. Hussain, M., Mousa Ali, E., Jarchavi, S. M. R., Zaidi, A., Najam, A. I., Alotaibi, A. A., & Ghoneim, S. S. (2022). Design and characterization of compact broadband antenna and its MIMO configuration for 28 GHz 5G applications. *Electronics, 11*(4), 523.
10. Khan, M. I., Khan, S., Kiani, S. H., Ojaroudi Parchin, N., Mahmood, K., Rafique, U., & Qadir, M. M. (2022). A compact mmWave MIMO antenna for future wireless networks. *Electronics, 11*(15), 2450.
11. Hussain, N., Awan, W. A., Ali, W., Naqvi, S. I., Zaidi, A., & Le, T. T. (2021). Compact wideband patch antenna and its MIMO configuration for 28 GHz applications. *AEU-International Journal of Electronics and Communications, 132*, 153612.
12. Karthikeya, G. S., Koul, S. K., Poddar, A. K., & Rohde, U. L. (2022). Compact bent-corner orthogonal beam switching antenna module for 5G mobile devices. *Journal of Electromagnetic Engineering and Science, 22*(1), 74–83.
13. Jeon, J., Lee, G., Ibrahim, A. A., Yuan, J., Xu, G., Cho, J., & Zhang, J. C. (2021). MIMO evolution toward 6G: Modular massive MIMO in low-frequency bands. *IEEE Communications Magazine, 59*(11), 52–58.
14. Li, J., Yue, D. W., & Sun, Y. (2018). Performance analysis of millimeter wave massive MIMO systems in centralized and distributed schemes. *IEEE Access, 6*, 75482–75494.
15. El-Nady, S. M., & Attiya, A. M. (2022). Periodically-stub-loaded microstrip line wideband circularly polarized millimeter wave MIMO antenna. *IEEE Access, 10*, 20465–20472.
16. Rosaline, I., Kumar, A., Upadhyay, P., & Murshed, A. H. (2022). Four element MIMO antenna systems with decoupling lines for high-speed 5G wireless data communication. *International Journal of Antennas and Propagation, 2022.*
17. Khan, M. A., Al Harbi, A. G., Kiani, S. H., Nordin, A. N., Munir, M. E., Saeed, S. I., & Dalarsson, M. (2022). mmWave Four-element MIMO antenna for future 5G systems. *Applied Sciences, 12*(9), 4280.
18. Bilal, M., Naqvi, S. I., Hussain, N., Amin, Y., & Kim, N. (2022). High-isolation MIMO antenna for 5G millimeter-wave communication systems. *Electronics, 11*(6), 962.
19. Patel, A., Vala, A., Desai, A., Elfergani, I., Mewada, H., Mahant, K., & Rodriguez, J. (2022). Inverted-L shaped wideband MIMO antenna for millimeter-wave 5G applications. *Electronics, 11*(9), 1387.

20. Hussain, R., Abou-Khousa, M., Iqbal, N., Algarni, A., Alhuwaimel, S. I., Zerguine, A., & Sharawi, M. S. (2022). A multiband shared aperture MIMO antenna for millimeter-wave and sub-6GHz 5G applications. *Sensors, 22*(5), 1808.
21. Askari, H., Hussain, N., Sufian, M. A., Lee, S. M., & Kim, N. (2022). A wideband circularly polarized magnetoelectric dipole antenna for 5G millimeter-wave communications. *Sensors, 22*(6), 2338.
22. Khan, J., Ullah, S., Ali, U., Tahir, F. A., Peter, I., & Matekovits, L. (2022). Design of a millimeter-wave MIMO antenna array for 5G communication terminals. *Sensors, 22*(7), 2768.
23. Alanazi, M. D., & Khamas, S. K. (2022). A compact dual band MIMO dielectric resonator antenna with improved performance for mm-Wave applications. *Sensors, 22*(13), 5056.
24. Sabek, A. R., Ali, W. A., & Ibrahim, A. A. (2022). Minimally coupled two-element MIMO antenna with dual band (28/38 GHz) for 5G wireless communications. *Journal of Infrared, Millimeter, and Terahertz Waves, 43*(3, 4), 335–348.
25. Nowak, T. W., Sepczuk, M., Kotulski, Z., Niewolski, W., Artych, R., Bocianiak, K., & Wary, J. P. (2021). Verticals in 5G MEC-use cases and security challenges. *IEEE Access, 9*, 87251–87298.

CHAPTER 10

Design of Multiband Antenna Using Flexible Substrate for Wireless and Sub-5G Applications

RAGHAV DWIVEDI,[1,*] RAJESH KUMAR DWIVEDI,[2] and VINOD KUMAR SINGH[3]

[1]*Department of Electronics and Communication Engineering, Pranveer Singh Institute of Engineering and Technology, Kanpur, Uttar Pradesh, India*

[2]*School of Basic Sciences, CSJMU, Kanpur, Uttar Pradesh, India*

[3]*S.R. Group of Institutions (CEMT), Jhansi, UP, India*

*Corresponding author

ABSTRACT

This chapter explores the design and optimization of antennas using flexible substrates for wireless communication and sub-5G applications. The evolution of wireless technologies necessitates innovative antenna designs with flexible substrates that offer advantages in conformability, lightweight properties, and integration ease. The study begins by reviewing current antenna design technology, highlighting conventional rigid substrate limitations and the opportunities presented by flexible alternatives. The proposed design leverages flexible substrates' mechanical properties for conformal integration with irregular surfaces crucial for modern communication devices. Advanced simulation tools and optimization techniques are employed to refine antenna geometry, focusing on key performance metrics, including impedance matching, radiation pattern, and bandwidth.

Wireless Technology: Advances and Innovations for Smart IoT Applications. Rajesh Kumar Dwivedi, Vinod Kumar Singh, and Raghav Dwivedi (Eds.)

DOI: 10.1201/9781998511495-10

The integration of flexible substrates presents unique challenges requiring a careful balance between mechanical flexibility and electrical performance. The research addresses strategies for managing flexible material integration challenges while ensuring design robustness and reliability. The study examines the substrate's impact on form factor and portability, which contribute to compact, lightweight device development for various deployment scenarios. This research advances wireless communication antenna design, particularly for sub-5G applications, demonstrating how flexible substrates can achieve enhanced adaptability and performance in contemporary and future communication systems.

10.1 INTRODUCTION

In the dynamic landscape of wireless communication technologies, the relentless pursuit of faster data rates, lower latency, and increased connectivity has fueled the need for innovative antenna designs [2]. This quest is further intensified with the advent of sub-5G applications, where the demand for versatile and high-performance communication systems has reached unprecedented levels [3, 4]. Traditional rigid substrates have long been the staple for antenna design; however, their limitations in conforming to irregular surfaces and adapting to the diverse requirements of modern applications have prompted a paradigm shift towards the integration of flexible substrates. This chapter embarks on a comprehensive exploration of the design intricacies and optimization strategies employed in antennas utilizing flexible substrates [5, 6]. The primary objective is to address the challenges posed by the evolving communication landscape, emphasizing the indispensability of flexible materials in meeting the stringent demands of wireless and sub-5G applications. The introduction provides a contextual backdrop by reviewing the current state-of-the-art in antenna design, shedding light on the limitations of conventional approaches, and setting the stage for the emergence of flexible substrates as a viable alternative [7, 8].

The escalating demand for ubiquitous connectivity and the proliferation of Internet of Things (IoT) devices underscore the need for antennas that are not only high-performing but also adaptable to diverse deployment scenarios [9]. This necessitates a departure from the rigidity of traditional substrates, paving the way for the exploration of materials that offer mechanical flexibility without compromising electrical performance [10]. The integration of flexible substrates in antenna design is envisioned as a transformative step towards achieving conformal and lightweight communication devices capable of seamlessly blending into their environment [11]. As wireless communication technologies evolve, the research community grapples with the complexities of designing antennas that can operate across a broad spectrum of frequencies, exhibit optimal impedance matching, and maintain stable radiation patterns [12]. The introduction sets the stage for the subsequent sections of this chapter, where advanced simulation tools and optimization techniques are employed to refine antenna geometries, ensuring superior performance metrics [13]. By synthesizing a comprehensive understanding of the challenges and opportunities inherent in antenna design, this research seeks to contribute to the ongoing discourse on the development of communication systems that transcend the limitations of conventional methodologies [14]. In the ensuing sections, the chapter delves into the intricate details of the antenna design process, highlighting key considerations such as impedance matching, radiation pattern optimization, and the integration of flexible substrates [15]. Through a systematic exploration of these elements, the research aims to provide valuable insights into the development of antennas tailored for the demands of wireless communication and the emerging era of sub-5G applications [16].

10.2 EXPERIMENTAL METHODS AND MATERIALS

10.2.1 JEANS AS FLEXIBLE SUBSTRATE

The choice of materials in antenna design plays a pivotal role in determining the overall performance and functionality of the system. In recent years, there has been a growing interest in utilizing unconventional materials to achieve lightweight, flexible, and conformal antenna structures. One such material that has gained attention is denim, a versatile fabric commonly associated with the fashion industry [17]. In this context, denim serves as a flexible substrate with a dielectric constant of 1.7 and a loss tangent of 0.024.

The dielectric constant, often denoted by the symbol εr, is a crucial parameter that characterizes how a material responds to an electric field. A lower dielectric constant, such as that of denim (1.7), implies reduced signal propagation speed within the material compared to traditional substrates with higher dielectric constants [18]. This characteristic makes denim particularly suitable for applications where impedance matching and control over the electromagnetic field are critical. Additionally, the loss tangent (tan δ) represents the ratio of the imaginary part to the real part of the dielectric constant and indicates the amount of energy absorbed by the material [19]. The low loss tangent of denim (0.024) suggests that it exhibits minimal energy dissipation, making it an attractive choice for antenna substrates where signal integrity and efficiency are paramount [20]. The unique combination of a moderate dielectric constant and low loss tangent in denim makes it an intriguing material for antenna design [21]. Its flexibility and conformability contribute to the development of antennas that can be seamlessly integrated into irregular or curved surfaces; a feature particularly valuable in modern communication devices. Furthermore, the use of denim as a substrate aligns with the growing trend toward sustainable and eco-friendly materials in various industries [22]. Repurposing denim for antenna applications not only taps into its electrical properties but also aligns with the principles of reusability and reduced environmental impact [23]. In summary, the selection of denim as an antenna substrate, with its dielectric constant of 1.7 and loss tangent of 0.024, represents a noteworthy exploration into unconventional materials for communication system design [24]. The unique electrical characteristics, coupled with the fabric's inherent flexibility and sustainability, position denim as a promising candidate for the development of antennas tailored for wireless and sub-5G applications [25]. Figure 10.1 shows the antenna design using the jeans substrate.

10.2.2 CST STUDIO SUITE AS ANTENNA DESIGN TOOL

CST Studio Suite is indeed a widely used software suite for electromagnetic simulation, including antenna design. CST Studio Suite, developed by Computer Simulation Technology (CST), is known for its capabilities in simulating and optimizing various electromagnetic devices and systems. It is commonly employed in the design and analysis of antennas, RF/microwave components, and other electromagnetic structures [26].

Key features of CST Studio Suite for antenna design include:

1. **Electromagnetic Simulation:** The software uses various numerical techniques, such as the Finite Integration Technique (FIT), to accurately simulate electromagnetic fields, ensuring a comprehensive analysis of antenna behavior.
2. **Antenna Design and Optimization:** CST Studio Suite provides tools for designing and optimizing antenna structures. Users can explore different configurations, materials, and geometries to achieve desired performance characteristics.

3. **Parametric Studies:** Engineers and researchers can conduct parametric studies to analyze how changes in parameters (such as dimensions or materials) impact the antenna's performance, helping in optimization.
4. **Integration with Other Technologies:** The suite is versatile and can be used for the design and simulation of various electromagnetic devices beyond antennas, including filters, couplers, and other RF/microwave components.
5. **Frequency Domain and Time Domain Simulations:** CST Studio Suite supports both frequency domain and time domain simulations, allowing users to analyze the antenna's behavior across different operating conditions.
6. **Wide Frequency Range:** It is suitable for simulations across a broad frequency range, making it applicable to a variety of antenna types used in different communication systems.
7. **Material Database:** The software often includes a comprehensive material database, allowing users to select from a range of predefined materials or input custom material properties.

Users can benefit from CST Studio Suite's capabilities in terms of accuracy, efficiency, and flexibility, making it a valuable tool in the field of antenna design and electromagnetic simulation. It is commonly utilized in both academic research and industrial applications for the development of advanced wireless communication systems.

FIGURE 10.1 Antenna design using the jeans substrate for reference.

Source: Reprinted with permission from [1]. http://creativecommons.org/licenses/by/4.0/

The antenna is designed using CST Studio Suite software, and the antenna was optimized using slotting techniques.

10.2.3 HIERARCHY OF OPTIMIZING PROPOSED ANTENNA

Designing a textile antenna involves integrating antenna components into fabric or textile materials. Textile antennas are commonly used in wearable technology, smart clothing, and other applications where flexibility and comfort are essential. Here are the general steps involved in designing a textile antenna:

1. **Define Requirements:** Determine the application and requirements of the antenna, such as frequency range, bandwidth, size, and flexibility.
2. **Select Antenna Type:** Choose the appropriate type of antenna for the application. Common textile antenna types include dipole antennas, microstrip antennas, and patch antennas.
3. **Select Textile Material:** Choose a suitable textile material that meets the design requirements. The material should be flexible, lightweight, and comfortable for the intended use.
4. **Conduct Material Characterization:** Understand the electromagnetic properties of the chosen textile material, such as permittivity and permeability, as these properties influence the antenna's performance.
5. **Design Antenna Geometry:** Determine the dimensions and geometry of the antenna based on the selected type and the desired operating frequency. Consider the textile material's impact on the antenna's performance.
6. **Simulate Antenna Performance:** Use electromagnetic simulation software to model and simulate the antenna's performance. Adjust the design parameters to optimize the antenna for the desired specifications.
7. **Prototype Fabrication:** Create a prototype of the textile antenna using the chosen textile material and the designed geometry.

Various manufacturing techniques, such as embroidery, printing, or weaving, may be employed depending on the material.

8. **Integration with Electronics:** Integrate the textile antenna with any necessary electronics, such as connectors, feedlines, and matching networks. Ensure that the connection points are secure and won't compromise the flexibility of the textile.
9. **Testing and Measurement:** Measure the actual performance of the textile antenna using specialized equipment. Compare the measured results with the simulation data to validate the design.
10. **Optimization:** If necessary, refine the design based on the test results. Make adjustments to improve performance or address any issues identified during testing.
11. **Durability Testing:** Evaluate the durability of the textile antenna by subjecting it to bending, washing, and other conditions it might encounter during use. Ensure that the antenna maintains its performance after repeated use and exposure to environmental factors.
12. **Finalization and Production:** Once the design meets the requirements and passes all tests, finalize the textile antenna design for production. Consider scalability and cost-effectiveness for mass production.

By following these steps, designers can create effective and reliable textile antennas suitable for various applications.

Within Figure 10.2 of our illustration, there is a nested orbicular-shaped structure that has been affixed with a radius identified as D_2, and its measured value registers at 18 mm. Following this, an additional ring is affixed onto the structure created in Figure 10.3(b), featuring an inner radius denoted as D_3, specifically set at 14 mm as depicted in Figure 10.3(c). Subsequently, this structure is subtracted from a circular structure with a radius of 18 mm, resulting in the formation of the structure showcased in Figure 10.3(d). Subsequent to this transformation, a diminutive structure is fashioned, possessing an area equivalent to $W_1 \times L_1$ mm^2, with both the length and width measured at 2×1.5 mm^2. The resultant antenna is visually presented in Figure 10.3(e). This evolving design is further enhanced by the introduction of another concentric cylinder with radii D_4 and D_5 set at 21 mm and 25 mm, respectively. This additional element is positioned atop the antenna depicted in Figure 10.3(f). Finally, to culminate our design process, a six-segment slot is meticulously carved into the middle, featuring a radius of 6 mm. This intricate step contributes to the realization of our ultimate proposed antenna, a depiction of which is presented in Figure 10.3(g). Finally, another slot is cut into the design in Figure 10.3(g) to get our proposed antenna. To provide a comprehensive overview of the geometric attributes and specifications of our proposed antenna, refer to the detailed information outlined in Table 10.1, corresponding to the visual representation in Figure 3.2. In the intricate tapestry of Figure 10.3's antenna evolution, each step unfolds as a carefully orchestrated dance between form and function. The initial integration of the nested orbicular structure, adorned with the defining radius D_2, acts as a cornerstone in the architectural symphony. This foundational radius, meticulously measured at 18 mm, propels the design journey forward, setting the stage for a cascade of creative enhancements. The subsequent introduction of an additional ring, with an inner radius D_3 meticulously set at 14 mm as captured in Figure 10.3(c), serves as a pivotal layer in the antenna's evolving narrative. This strategic augmentation not only refines the overall form but also introduces an interplay of geometries that lay the groundwork for subsequent design intricacies. The subtractive process, wherein the structure is judiciously removed from a circular counterpart with a radius of 18 mm, marks a transformative turning point. The resulting structure, unveiled in Figure 10.3(d), bears the imprints of these calculated alterations, embodying a synthesis of creative design and engineering precision. As the design narrative unfolds, a microcosm takes shape—a small yet pivotal structure is meticulously crafted, boasting an area precisely proportional to $W_1 \times L_1$ mm^2. The dimensions of 2×1.5 mm^2, carefully measured and integrated, contribute to the nuanced intricacies of the evolving antenna, as vividly portrayed in Figure 10.3(e). Continuing this artistic trajectory, a concentric cylinder with radii D_4 and D_5, thoughtfully configured at 21 mm and 25 mm, respectively, makes its entrance. This additional layer, elegantly mounted over the antenna structure of Figure 10.3(f), introduces a dimensional richness that adds depth and sophistication to the overall design. The crescendo of creativity culminates in the final act—a meticulous incision, a six-segment slot carved at the antenna's core, with a radius of 6 mm. This deliberate addition serves as the defining touch, transforming the amalgamation of shapes and structures into the ultimate proposed antenna, a visual spectacle embodied in Figure 10.3(g).

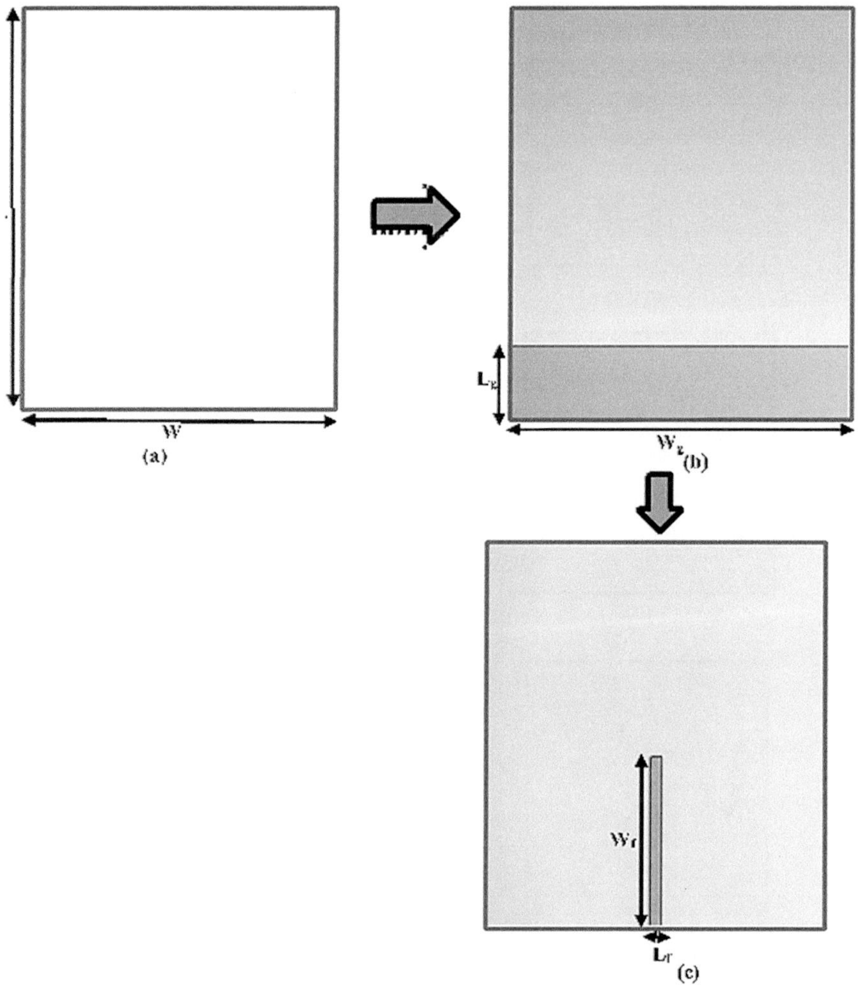

FIGURE 10.2 Partial ground of proposed antenna structure.

In this symphony of design, each element plays a harmonious role, contributing to the synthesis of a textile antenna that seamlessly melds artistry with technological innovation.

In the context of the evolving narrative, the significance of Figure 10.2 is heightened by the revelation that it specifically illustrates the partial ground configuration of the proposed antenna. This nuanced detail adds another layer of complexity and purpose to the design process.

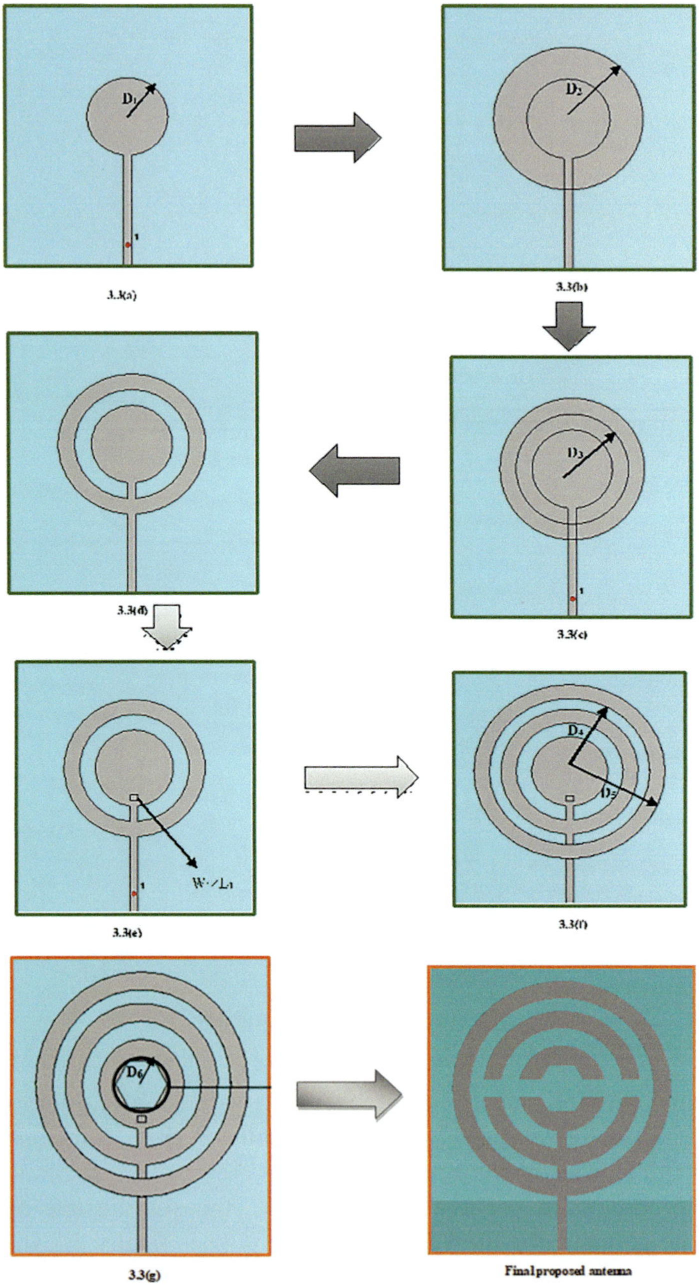

FIGURE 10.3 Hierarchy of optimizing the proposed antenna.

TABLE 10.1 Parameter of Proposed Antenna

Parameter	Value(mm)
W (width of substrate)	60
L (Length of substrate)	66
W_g (width of ground)	60
L_g (Length of ground)	12
W_f(Width of feed)	2
L_f(Length of feed)	29
D_1	10
D_2	18
D_3	14
D_4	21
D_5	25
W_1	2
L_1	1.5
D_6(with segment)	6(6 segment)
h (height of substrate)	1
ε_r	1.7(jeans)
Loss tangent	0.024

The inclusion of a partial ground in the antenna's architecture serves as a deliberate choice, influencing the electromagnetic characteristics and performance of the overall structure.

As we delve into the visual representation of Figure 10.3, the intricacies of the proposed antenna's partial ground become apparent. The nested orbicular structure, the carefully added ring, and the subsequent subtractive process all contribute to the strategic shaping of this partial ground. This feature plays a crucial role in enhancing the antenna's efficiency, radiation pattern, and resonance characteristics. The delineation of a small yet pivotal structure, meticulously designed with specific dimensions and proportions, further underscores the precision involved in incorporating the partial ground. This microcosm, as highlighted in Figure 10.3(e), is not merely an isolated design element but an integral part of the overall antenna structure, intricately contributing to the shaping of its partial ground. The concentric cylinder with radii D_4 and D_5, as portrayed in Figure 10.3(f), not only adds physical layers to the antenna but also influences the nature of the partial ground, introducing additional complexities and interactions within the electromagnetic domain. The final touch, represented by the six-segment slot with a radius of 6 mm in Figure 10.3(g), further refines the partial ground configuration. This deliberate slotting contributes to the overall impedance matching and tuning of the antenna, ensuring that it aligns seamlessly with the desired operational frequencies. In summary, Figure 10.3 emerges as a visual exposition of the carefully crafted partial ground of the proposed antenna. Each design element, meticulously laid out and intricately interconnected, serves a purpose in shaping the antenna's electromagnetic characteristics, making it a testament to the fusion of artistic design and scientific precision in the realm of textile antenna engineering.

The proposed antenna boasts a carefully considered set of parameters, each contributing to its overall design and functionality. The substrate, characterized by a width (W) of 60 units and a length (L) of 66 units, establishes the foundational dimensions of the antenna. The ground plane, featuring a width (W_g) matching the substrate at 60 units and a length (L_g) of 12 units, serves as a crucial element influencing the antenna's electromagnetic characteristics. The feed structure, with a narrow width (W_f) of 2 units and a length (L_f) extending to 29 units, plays a pivotal role in delivering the input signal to the antenna. The circular structures, denoted by D_1, D_2, D_3, D_4, and D_5, showcase diameters of 10, 18, 14, 21, and 25 units, respectively, contributing to the intricate shaping of the antenna. Additionally, the specific geometric dimensions of a small structure ($W_1 \times L_1$) designed with dimensions 2×1.5 units add a microcosmic layer to the overall antenna design. The inclusion of a six-segment slot, with a radius (D_6) of 6 units, further refines the antenna's structure, influencing its resonance and impedance characteristics. The substrate's height (h) is set at 1 unit, creating a thin and flexible foundation for the antenna. The choice of material is marked by an ε_r (relative permittivity) value of 1.7, reflective of a jeans fabric. Moreover, the antenna accounts for a loss tangent of 0.024, considering the material's dissipative characteristics.

In essence, the proposed antenna parameters form a harmonious ensemble, meticulously tuned to achieve optimal performance. The interplay of substrate dimensions, ground characteristics, feed structure, and the strategic arrangement of circular and segment structures all contribute to the antenna's effectiveness, making it a well-rounded and purposeful design for its intended application. The following proposed antenna results have been discussed in the next paragraph.

10.3 RESULTS AND DISCUSSION

10.3.1 RETURN LOSS (RL) OF PROPOSED ANTENNA

The RL of an antenna is a measure of how well the antenna impedance matches the transmission line or system it is connected to. It is usually expressed in decibels (dB) and provides insight into the efficiency of power transfer between the antenna and the connected system. The formula for Return Loss is given by:

$$Rl(dB) = 20log_{10}\left[\frac{V_{reflection}}{V_{Incidence}}\right] \quad (10.1)$$

where, $V_{reflection}$ = Amplitude of the reflected wave; $V_{Incidence}$ = Amplitude of the incidence wave.

To determine the Return Loss of the proposed antenna, you would need to simulate or measure the antenna's impedance characteristics at the desired operating frequency. The reflection coefficient, which is the ratio of reflected voltage to incident voltage, can be derived from the impedance. Given the specific geometric parameters of the proposed antenna (substrate dimensions, ground dimensions, feed structure, circular structures, etc.), and considering the material properties (ε_r and loss tangent), you would use electromagnetic simulation software or measurement equipment to obtain the impedance at the operating frequency of interest. Once the reflection coefficient is known, you can apply the formula above to calculate the Return Loss in decibels. Figure 10.4 shows the Return Loss of the proposed antenna.

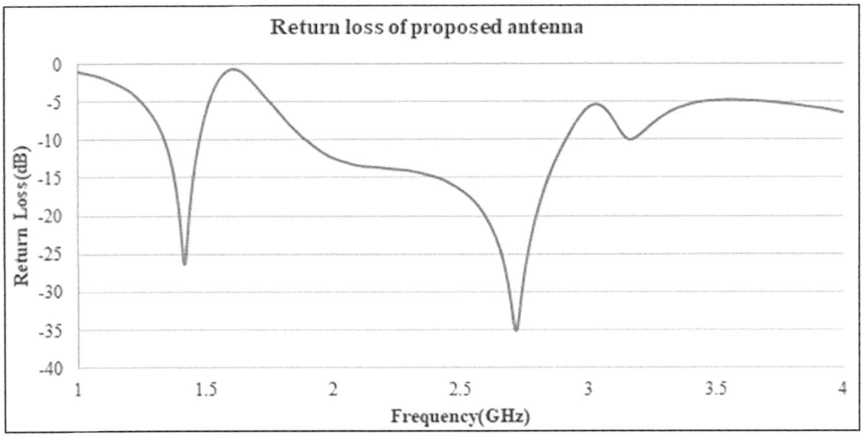

FIGURE 10.4 Return loss of the proposed antenna.

A higher return loss value indicates better impedance matching and, consequently, more efficient power transfer. It's important to note that the specific numerical value of the return loss would depend on the antenna's performance at the chosen operating frequency and the accuracy of the simulation or measurement data.

10.3.2 GAIN OF THE PROPOSED ANTENNA

The gain of an antenna is a measure of its directional ability to focus or concentrate the radiated energy in a particular direction. It is often expressed in decibels (dBi) and represents the ratio of the power radiated in the direction of interest to the power that would be radiated by an ideal isotropic radiator. The formula for antenna gain (G) is given by Equation 10.2.

$$G(dBi) = 10.\log_{10} \frac{P_{rad.main}}{P_{rad.iso}} \tag{10.2}$$

where, $P_{radmain}$= Power radiated in the main direction of interest; $P_{rad,\ iso}$= Power radiated by an isotropic radiator (an ideal point source radiating equally in all directions).

To determine the gain of the proposed antenna, you would typically use simulation software or measurement equipment to analyze the antenna's radiation pattern. The radiation pattern describes how the antenna radiates energy in three-dimensional space. After obtaining the radiation pattern, you can determine the power radiated in the main direction and calculate the gain using the formula above. It's important to note that the gain can vary depending on the frequency of operation, and the antenna is usually characterized at its resonant frequency or the frequency of interest. In summary, to find the gain of the proposed antenna, you need to analyze its radiation pattern at the desired operating frequency and apply the gain formula using the power radiated in the main direction of interest and the power radiated by an isotropic radiator. The proposed antenna has a maximum gain of 4.877 dB, as shown in Figure 10.5.

10.3.3 RADIATION PATTERN OF PROPOSED ANTENNA

The radiation pattern of the proposed antenna offers a comprehensive insight into its directional characteristics and performance. Through meticulous

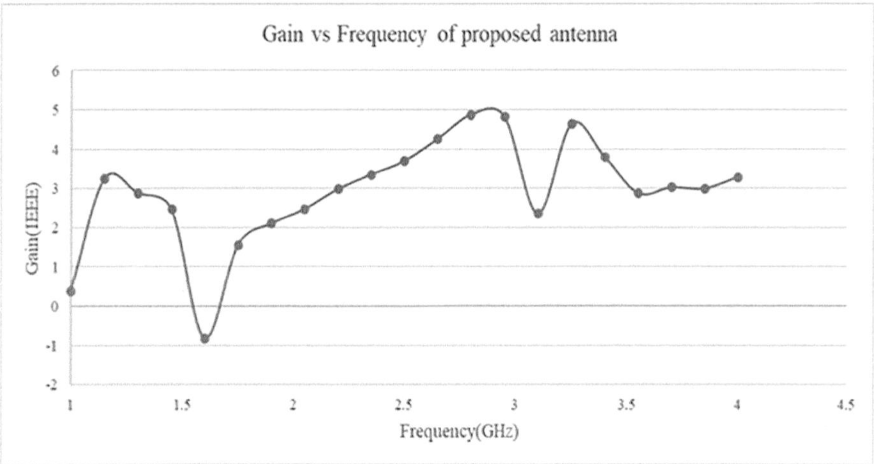

FIGURE 10.5 Gain of the proposed antenna.

analysis utilizing electromagnetic simulation software or precise measurement techniques, the antenna's radiation pattern is elucidated, providing a visual representation of how it distributes electromagnetic energy in three-dimensional space. This pattern showcases the antenna's ability to focus or concentrate radiation in specific directions, as well as the extent of its coverage across different angles. Key parameters such as directivity, beamwidth, and sidelobe levels are discerned from the radiation pattern, offering critical information about the antenna's effectiveness in transmitting or receiving signals within its operational frequency range. By examining the radiation pattern, engineers and researchers can assess the antenna's suitability for various applications, ensuring optimal performance and alignment with specific requirements. Additionally, insights gleaned from the radiation pattern aid in fine-tuning the antenna's design parameters to achieve desired performance metrics, further enhancing its functionality and versatility in real-world scenarios. The radiation of the proposed antenna is shown in Figure 10.6.

10.3.4 CURRENT DISTRIBUTION OF THE PROPOSED ANTENNA

The current distribution of the proposed antenna provides crucial insights into its electromagnetic behavior and functionality. Through electromagnetic simulation or measurement techniques, the distribution of electrical

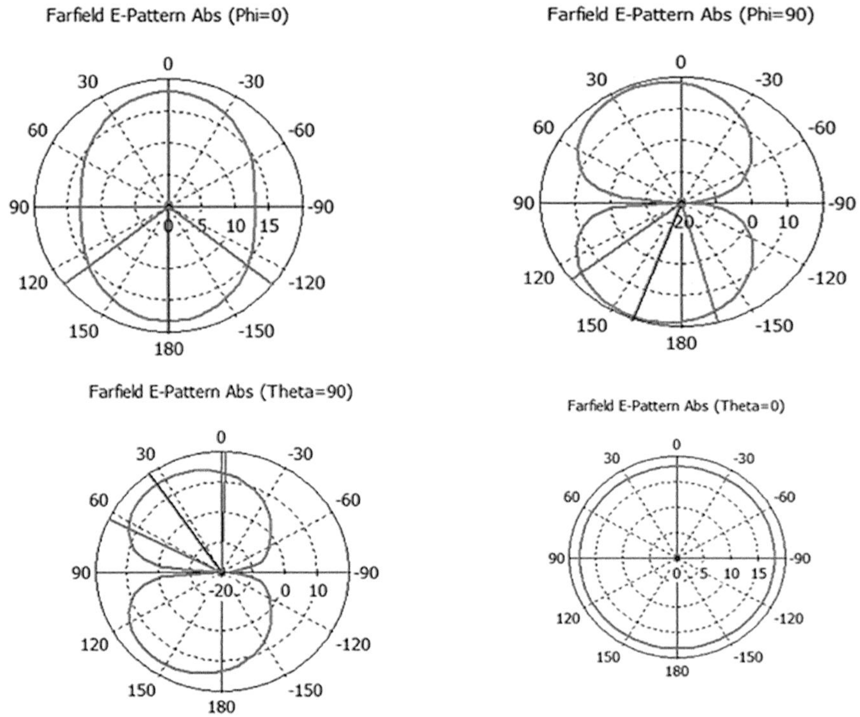

FIGURE 10.6 Radiation pattern of the proposed antenna.

current along the antenna's structure is visualized and analyzed. This distribution reveals how the antenna efficiently converts input electrical signals into radiated electromagnetic waves. By examining the current distribution, engineers can understand how energy flows through different segments of the antenna, identifying areas of high and low current density. This information is essential for optimizing the antenna's design parameters, such as dimensions, material properties, and feed configuration, to achieve desired performance characteristics. Moreover, the current distribution aids in identifying potential areas of impedance mismatch or energy loss, allowing for targeted adjustments to enhance the antenna's efficiency and effectiveness. Overall, a thorough understanding of the current distribution is instrumental in refining the proposed antenna design to meet specific application requirements and ensure optimal performance in real-world scenarios. The current distribution of the proposed antenna is shown in Figure 10.7.

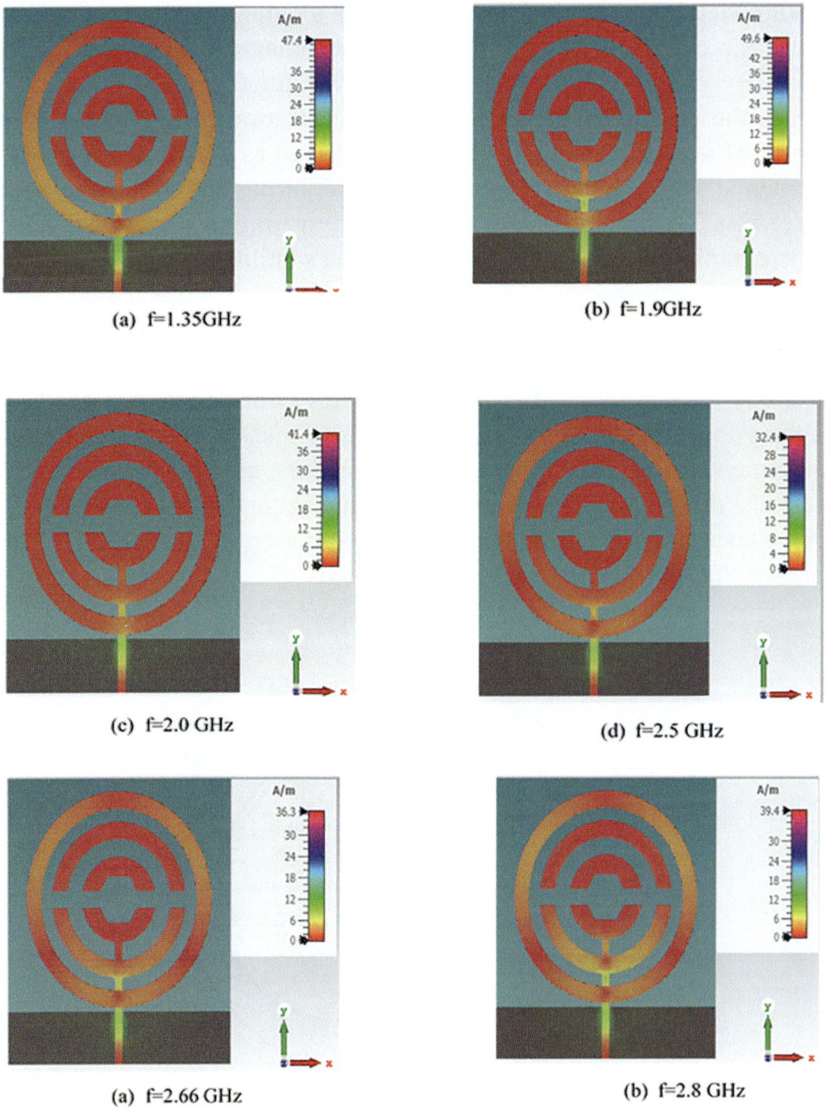

FIGURE 10.7 Current distribution of the proposed antenna.

10.3.5 DIRECTIVITY OF THE PROPOSED ANTENNA

The directivity of the proposed antenna is a pivotal metric that elucidates its efficiency in concentrating radiated energy towards specific directions.

Through meticulous analysis of the antenna's radiation pattern, engineers discern the extent to which it focuses electromagnetic waves at desired angles. By characterizing the peak gain within the main direction of interest, the directivity quantifies this concentration of radiation relative to an ideal isotropic radiator. This metric serves as a critical indicator of the antenna's effectiveness in applications requiring targeted communication or sensing capabilities. A high directivity implies a greater ability to focus radiation, maximizing the antenna's utility in scenarios where precise signal transmission or reception is paramount. Conversely, a lower directivity may indicate broader radiation coverage, suitable for applications demanding omnidirectional communication. Understanding the directivity of the proposed antenna facilitates informed design decisions, enabling engineers to tailor its parameters to meet specific performance requirements. Moreover, it guides optimizations aimed at enhancing the antenna's directional characteristics, ensuring optimal performance in diverse operational environments. The directivity of the proposed antenna is shown in Figure 10.8.

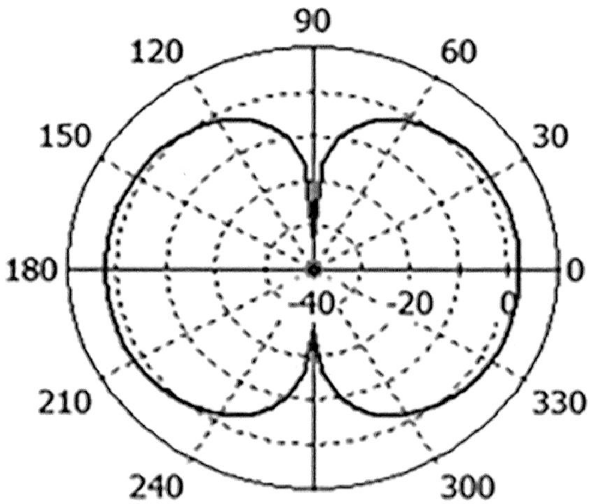

FIGURE 10.8 Directivity of the proposed antenna.

So, the proposed antenna has two bands centered at f = 1.417 GHz, having a return loss of –26.334 dB, and the second band is at f = 2.719 GHz, having a return loss of –35.075 dB, as shown in Figure 10.9.

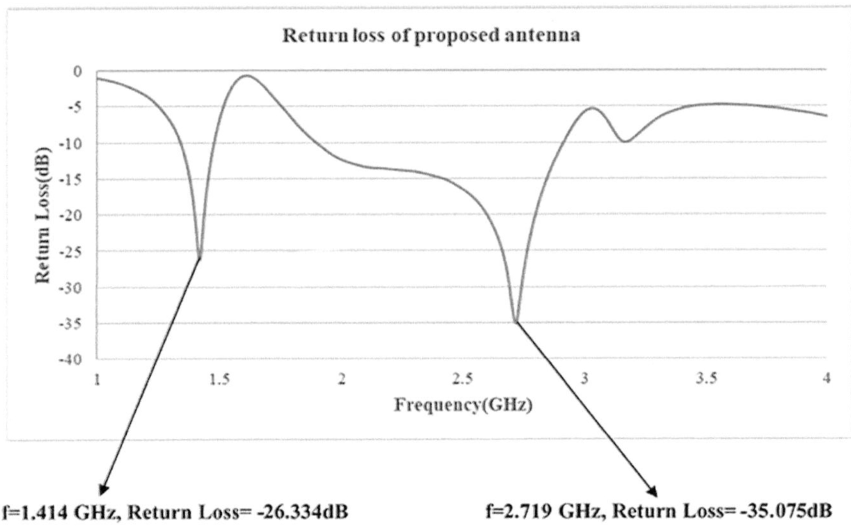

FIGURE 10.9 Multiband of the proposed antenna.

Dual-band antennas are increasingly sought after in wireless communication systems due to their ability to operate across multiple frequency bands, enabling versatile functionality and improved spectrum utilization. The proposed antenna design presents a promising solution with resonance frequencies centered at 1.417 GHz and 2.719 GHz, catering to the demands of modern communication systems. This chapter aims to delve into the return loss characteristics of the antenna within these frequency bands, shedding light on the impedance matching properties and overall performance. The analysis begins with a detailed examination of the return loss performance of the proposed antenna at the two resonance frequencies. At 1.417 GHz, the return loss is measured at –26.334 dB, indicating a favorable impedance match between the antenna and the transmission line. This value suggests minimal power reflection and efficient power transfer at this frequency, enhancing the antenna's performance in the corresponding band. Similarly, at 2.719 GHz, the return loss value of –35.075 dB signifies an even more robust impedance match, further optimizing the antenna's performance in the second frequency band. These return loss values are indicative of the antenna's ability to effectively radiate energy within the desired frequency ranges, minimizing losses and maximizing signal reception or transmission. The observed return loss values hold significant implications for the proposed antenna design.

The high return loss values at both resonance frequencies underscore the effectiveness of the antenna's impedance matching techniques, ensuring minimal signal reflection and improved transmission efficiency. This is particularly crucial in dual-band applications where a seamless transition between frequency bands is required. Additionally, the precise tuning of the antenna's resonance frequencies contributes to its versatility and adaptability in diverse operating environments. The implications of these return loss values extend beyond basic performance metrics, highlighting the antenna's potential for use in various communication systems, including wireless networks, satellite communication, and IoT devices.

10.4 CONCLUSION

In conclusion, the analysis of the proposed antenna design has provided valuable insights into its dual-band characteristics and impedance matching performance. With resonance frequencies centered at 1.417 GHz and 2.719 GHz, accompanied by return loss values of –26.334 dB and –35.075 dB respectively, the antenna demonstrates efficient power transfer and optimal impedance matching within the desired frequency bands. Additionally, the maximum gain of 4.877 dB underscores the antenna's ability to concentrate radiated energy in specific directions, further enhancing its performance. Looking ahead, there are several avenues for future research and development to explore. Firstly, further optimization of the antenna's performance could be pursued through iterative design refinements, such as fine-tuning geometric parameters and material properties to achieve even higher return loss values and gain levels. Additionally, investigations into enhancing the antenna's bandwidth and frequency agility could broaden its applicability across a wider range of communication systems and frequency bands. Moreover, the proposed antenna design lays the foundation for exploring advanced antenna technologies, such as phased array antennas or reconfigurable antennas, which offer enhanced versatility and adaptability to changing operational requirements. These technologies could be integrated into future iterations of the antenna design to unlock new capabilities and functionalities. Furthermore, real-world validation and testing of the proposed antenna in practical scenarios would provide valuable validation of its performance and reliability. Field trials and deployment in operational environments could offer insights into the antenna's robustness, durability, and effectiveness in real-world applications.

In conclusion, the proposed antenna design holds promise for various communication systems and wireless applications, with its dual-band characteristics, efficient impedance matching, and respectable gain levels. Continued research and development efforts in optimizing its performance and exploring advanced antenna technologies will further elevate its potential and pave the way for advancements in wireless communication and connectivity. Beyond its impressive performance metrics, the proposed antenna design holds significant promise for a wide array of practical applications across diverse fields.

1. **Wireless Communication Systems:** The dual-band characteristics of the antenna make it well-suited for use in wireless communication systems, including cellular networks, Wi-Fi routers, and Bluetooth devices. Its ability to operate efficiently across multiple frequency bands enhances network flexibility and reliability, ensuring seamless connectivity in dynamic environments.

2. **Satellite Communication:** The antenna's high gain and dual-band capabilities make it a suitable candidate for satellite communication systems. It can facilitate reliable data transmission between ground stations and satellites, supporting various applications such as remote sensing, weather monitoring, and global communication networks.

3. **Internet of Things (IoT) Devices:** In the burgeoning IoT ecosystem, where connectivity and energy efficiency are paramount, the proposed antenna can play a crucial role. Its compact size, dual-band operation, and efficient power transfer make it ideal for integration into IoT devices such as smart sensors, wearables, and connected appliances, enabling seamless communication and data exchange.

4. **Radar Systems:** These require antennas with high gain and precise beam steering capabilities for accurate detection and tracking. The proposed antenna's maximum gain and directional characteristics make it well-suited for radar applications, including air traffic control, weather monitoring, and military surveillance.

5. **Wireless Sensor Networks:** In applications where distributed sensing and data collection are essential, such as environmental monitoring, industrial automation, and smart infrastructure, the proposed antenna can serve as a reliable communication link. Its dual-band operation allows for versatile deployment in diverse

environments, while its efficient power transfer ensures long-term operation and minimal energy consumption.

6. **5G Networks:** As the deployment of 5G networks continues to expand, there is a growing demand for antennas that can support the high data rates and low latency requirements of next-generation communication systems. The proposed antenna's dual-band capabilities and high gain make it well-suited for use in 5G base stations, small cells, and other infrastructure components, facilitating the delivery of high-speed, low-latency connectivity to users worldwide.

In summary, the proposed antenna design holds immense potential for a wide range of applications, spanning from wireless communication and satellite systems to IoT devices and radar technology. Its impressive performance metrics, including dual-band operation, high gain, and efficient power transfer, position it as a versatile and reliable solution for addressing the connectivity needs of diverse industries and emerging technologies. Continued research and development efforts in optimizing its performance and exploring new application domains will further solidify its position as a key enabler of future wireless networks and communication systems.

KEYWORDS

- **flexible substrate**
- **IoT application**
- **multiband antenna**
- **return loss**
- **sub-5G applications**
- **wireless application**
- **wireless communication technologies**

REFERENCES

1. Yadav, A., Singh, V. K., Yadav, P., Beliya, A. K., Bhoi, A. K., & Barsocchi, P. (2020). Design of circularly polarized triple-band wearable textile antenna with safe low SAR for human health. *Electronics, 9*(9), 1366.

2. Sharma, N., & Bhatia, S. S. (2020). Performance enhancement of nested hexagonal ring-shaped compact multiband integrated wideband fractal antennas for wireless applications. *International Journal of RF and Microwave Computer-Aided Engineering, 30*(3), e22079.
3. Kumar, T. M., Jonathan, N. R., Peshwe, P., Doddipalli, S., & Kothari, A. (2020). An Annular Ring Antenna with Slotted Ground Plane for Dual Band Wireless Applications. In *Advances in Decision Sciences, Image Processing, Security and Computer Vision: International Conference on Emerging Trends in Engineering (ICETE), Vol. 2* (pp. 307–313). Springer International Publishing.
4. Peshwe, P., Doddipalli, S., & Kothari, A. (2019). An Annular Ring Antenna with Slotted Ground Plane for Dual Band Wireless Applications. In *Advances in Decision Sciences, Image Processing, Security and Computer Vision: International Conference on Emerging Trends in Engineering (ICETE), Vol. 2* (Vol. 4, p. 307). Springer.
5. Ketavath, K. N. (2019). Enhancement of gain with coplanar concentric ring patch antenna. *Wireless Personal Communications, 108*(3), 1447–1457.
6. Singh, V. K., & Bangari, S. D. N. (2017). Wearable Ultra-wide Dual Band Flexible Textile Antenna for WiMax/WLAN Application. *Wireless Personal Communications, 95*, 557–564.
7. Sayidmarie, K., & Nagem, T. A. (2012). Compact dual-band dual-ring printed monopole antennas for WLAN applications. *Progress In Electromagnetics Research B, 43*, 313–331.
8. Munde, M., Nandgaonkar, A., & Deosarkar, S. (2020). Dual feed wideband annular ring microstrip antenna with circular DGS for reduced SAR. *Progress In Electromagnetics Research B, 88*, 175–195.
9. George, N. M., Pushpa, T. A. J. M., & Mary, J. (2022). Durable silicon rubber-based miniaturized antenna with concentric circle structure for a medical telemetry application. *Progress In Electromagnetics Research M, 107*, 155–165.
10. Boutejdar, A., & Halim, B. I. (2019). Design and manufacturing of a dual patch ring antenna using parasitic ring resonators and partial ground plane for multiband applications. In *2019 IEEE International Symposium on Phased Array System & Technology (PAST)* (pp. 8138–8143). IEEE.
11. Xu, L. J., Bo, Y., Lu, W. J., Zhu, L., & Guo, C. F. (2019). Circularly polarized annular ring antenna with wide axial-ratio bandwidth for biomedical applications. *IEEE Access, 7*, 59999–60009.
12. Wa'il, A., Shaaban, R. M., & Duffy, A. P. (2022). Design, simulation, and fabrication of a double annular ring microstrip antenna based on gaps with multiband feature. *Engineering Science and Technology, an International Journal, 29*, 101033.
13. Kim, J. H., Woo, D. W., & Park, W. S. (2009). Design of a dual-mode annular ring antenna with a coupling feed. *Microwave and Optical Technology Letters, 51*(12), 3011–3013.
14. Le, T. T., Kim, Y. D., & Yun, T. Y. (2021). A triple-band dual-open-ring high-gain high-efficiency antenna for wearable applications. *IEEE Access, 9*, 118435–118442.
15. Akalya, C. G., Kumar, D. S., & Rao, P. H. (2021). Design and Simulation of Flexible Antenna with a Defected Ground Structure for Wireless Applications. In *International Conference on Futuristic Communication and Network Technologies* (pp. 137–152). Singapore: Springer Nature Singapore.

16. Sankaran, K. S. (2023). Design of Reconfigurable Antenna with Defected Ground Structure for Wireless Sensor Network. *Wireless Personal Communications, 129*(3), 1771–1788.
17. Zainudin, N. A. F. M., Osman, M. N., Sabapathy, T., & Jusoh, M. (2020). Investigation on the effect of deflected ground structure in multi-direction steerable antenna for WSN. In *IOP Conference Series: Materials Science and Engineering* (Vol. 767, No. 1, p. 012024). IOP Publishing.
18. Abutarboush, H. F., Li, W., & Shamim, A. (2020). Flexible-screen-printed antenna with enhanced bandwidth by employing defected ground structure. *IEEE Antennas and Wireless Propagation Letters, 19*(10), 1803–1807.
19. Khatoon, S., Marwah, N., & Akhtar, J. (2022). Design of Flexible Antenna with Defected Ground Structure. In *International Conference on Communication and Intelligent Systems* (pp. 541–551). Singapore: Springer Nature Singapore.
20. Gupta, S. K., & Bage, A. (2020). A compact, T-slotted wide band microstrip antenna with defected ground structure for future 5G communications. In *2020 URSI Regional Conference on Radio Science (URSI-RCRS)* (pp. 1–4). IEEE.
21. Hossain, A. R., Mertvyy, A. A., Tran, N., & Karacolak, T. (2022). A high gain flexible antenna for full duplex system at 5.8 GHz with defected ground structure. In *2022 IEEE International Symposium on Antennas and Propagation and USNC-URSI Radio Science Meeting (AP-S/URSI)* (pp. 1040–1041). IEEE.
22. Dwivedi, R., Srivastava, D. K., & Singh, V. K. (2023). A novel miniaturized triangular slotted and DGS based inverted circular key shaped textile antenna with enhanced bandwidth for C, WLAN, RAS, N795G band application. *Iranian Journal of Science and Technology, Transactions of Electrical Engineering, 48*, 17–36.
23. Dhupkariya, S., & Singh, V. K. (2015). Textile antenna for C-band satellite communication application. *Journal of Telecommunication, Switching Systems and Networks, 2*(2), 20–25.
24. Memon, A. W., Van Torre, P., Malengier, B., & Van Langenhove, L. (2020). Investigating the effect of air permeability and moisture on 2.5 GHz textile microstrip patch antenna performance. In *IOP Conference Series: Materials Science and Engineering* (Vol. 827, No. 1, p. 012048). IOP Publishing.
25. El Gharbi, M., Martinez-Estrada, M., Fernández-García, R., Ahyoud, S., & Gil, I. (2021). A novel ultra-wide band wearable antenna under different bending conditions for electronic-textile applications. *The Journal of The Textile Institute, 112*(3), 437–443.
26. Kao, H. L., & Chuang, C. H. (2020). Bending effects on a fabric-based antenna for wearable applications. In *2020 IEEE 70th Electronic Components and Technology Conference (ECTC)* (pp. 1665–1670). IEEE.
27. Ullah, I., Wagih, M., & Beeby, S. P. (2022). Design of textile antenna for moisture sensing. *Engineering Proceedings, 15*(1), 11.

CHAPTER 11

Design and Fabrication of Square-Shaped Slot-Cut Microstrip Patch Antenna for IoT Applications

SURJEET RAIKWAR[1], RAGHAV DWIVEDI,[2*] and D. K. SRIVASTAVA[1]

[1]*ECE Department, Bundelkhand Institute of Engineering and Technology, Jhansi, Uttar Pradesh, India*

[2]*Department of Electronics and Communication Engineering, Pranveer Singh Institute of Engineering and Technology, Kanpur, Uttar Pradesh, India*

*Corresponding author

ABSTRACT

This chapter presents the design and fabrication of a dual-band microstrip patch antenna with dimensions of 20 mm by 20 mm. The proposed antenna incorporates a slot cut measuring 4 mm by 4 mm to achieve the desired dual-band functionality. The performance of the antenna is evaluated using several key metrics, including return loss, fractional bandwidth, gain, voltage standing wave ratio (VSWR), and directivity. These parameters provide insights into the antenna's impedance matching, bandwidth coverage, radiation efficiency, and directional characteristics. Through rigorous analysis and experimentation, the chapter aims to assess the suitability of the proposed dual-band microstrip patch antenna for various wireless communication applications. The findings contribute to the ongoing advancements in antenna design and technology, offering valuable insights into the performance characteristics and potential practical applications of dual-band microstrip patch antennas in modern communication systems.

Wireless Technology: Advances and Innovations for Smart IoT Applications. Rajesh Kumar Dwivedi, Vinod Kumar Singh, and Raghav Dwivedi (Eds.)

DOI: 10.1201/9781998511495-11

11.1 INTRODUCTION

The advent of wireless communication technologies has revolutionized the way we connect and interact with the world around us. In this era of ubiquitous connectivity, the demand for compact, efficient, and versatile antennas has never been greater [1]. Microstrip patch antennas have emerged as a popular choice for various wireless communication applications due to their low profile, lightweight, ease of fabrication, and compatibility with integrated circuitry [2]. This chapter focuses on the design and fabrication of a dual-band microstrip patch antenna, a key component in modern communication systems [3]. With dimensions of 20 mm by 20 mm, the proposed antenna is compact yet capable of operating across two distinct frequency bands [4]. The incorporation of a 4 mm by 4 mm slot cut enhances the antenna's dual-band functionality, allowing it to meet the evolving requirements of wireless communication systems [5].

The performance evaluation of the antenna is conducted using several essential metrics, including return loss, fractional bandwidth, gain, voltage standing wave ratio (VSWR), and directivity [6]. These parameters provide valuable insights into the antenna's impedance matching, bandwidth coverage, radiation efficiency, and directional characteristics [7]. By analyzing these performance metrics, we can assess the suitability of the proposed dual-band microstrip patch antenna for a wide range of wireless communication applications [8]. Through rigorous analysis and experimentation, this chapter aims to contribute to the ongoing advancements in antenna design and technology [9]. By exploring the design considerations, fabrication techniques, and performance characteristics of the dual-band microstrip patch antenna, we seek to enhance our understanding of its capabilities and potential applications in modern communication systems [9]. Ultimately, the insights gained from this study may pave the way for the development of more efficient, reliable, and versatile antennas to meet the ever-growing demands of wireless communication technologies [10].

The Internet of Things (IoT) has revolutionized the way devices communicate and interact with each other, ushering in an era of interconnected smart devices [11]. As IoT applications continue to proliferate across various industries, the demand for compact, efficient, and reliable antennas tailored for IoT devices has surged [12]. In response to this need, this study focuses on the design and fabrication of a square-shaped slot cut microstrip patch antenna specifically optimized for IoT applications [13]. The design of the square-shaped slot cut microstrip patch antenna is meticulously tailored to meet the unique requirements of IoT applications [14]. The antenna's compact size and square shape make it ideal for integration into small IoT devices while maintaining optimal performance [15]. The incorporation of a slot cut further enhances the antenna's functionality by enabling dual-band operation or enhanced bandwidth, essential for accommodating diverse IoT communication protocols and frequency bands [16]. The fabrication process involves precise manufacturing techniques to ensure the antenna's structural integrity and electromagnetic performance.

High-quality substrate materials, such as Rogers or FR-4, are chosen to provide the necessary dielectric properties and mechanical stability [17]. The conductive patch and feeding structure are meticulously etched onto the substrate using photolithography or printed circuit board (PCB) manufacturing processes. The slot cut is carefully integrated into the design to achieve the desired dual-band or broadband characteristics, enhancing the antenna's versatility for IoT applications [18]. The performance of the square-shaped slot cut microstrip patch antenna is rigorously evaluated using various metrics, including return loss, bandwidth, gain, radiation pattern, and efficiency [19]. These metrics provide valuable insights into the antenna's impedance matching, frequency coverage, radiation characteristics, and overall performance suitability for IoT applications [20]. Through comprehensive testing and analysis, the antenna's ability to meet the stringent requirements of IoT communication systems is assessed, ensuring optimal performance and reliability in real-world deployment scenarios [21].

In conclusion, the design and fabrication of the square-shaped slot cut microstrip patch antenna represent a significant advancement in antenna technology tailored for IoT applications [22]. By combining compact size, dual-band capability, and efficient fabrication techniques, this antenna offers a versatile solution for integrating wireless connectivity into IoT devices [23]. As IoT continues to evolve and expand into new domains, the square-shaped slot cut microstrip patch antenna stands poised to play a pivotal role in enabling seamless and reliable communication between interconnected smart devices [24].

11.2 ANTENNA DESIGN FORMULATION

The length and width of the microstrip antenna can be found out using the Equation 11.1

$$W = \frac{c}{fr}\sqrt{\frac{2}{\varepsilon_r + 1}} \tag{11.1}$$

Where f_r = design frequency (in GHz), εr is the dielectric constant of substrate (FR4=4.4), c is speed of light in vacuum 3 × 10^8 m/s, and W is the width of patch

When, $\frac{W_p}{h} > 1$, then the Z_o *and* $\in_{reff}$ can be calculated by using Equations 11.2 and 11.3.

$$Z_o = Z_f \; 2\pi\sqrt{\varepsilon_{reff}}\, ln(\, 8\,\frac{h}{Wp} + \frac{Wp}{h}) \tag{11.2}$$

$$\varepsilon_{reff} = \frac{\varepsilon_r + 1}{2} + \left[\frac{\varepsilon_r - 1}{2}\left(1 + 12\frac{h}{W_p}\right)^{-\frac{1}{2}} + 0.04(1 - \frac{W_p}{h}^{2}\right] \tag{11.3}$$

When, $\frac{W_p}{h} > 1$, then the Z_o *and* $\in_{reff}$ can be calculated by using Equations 11.4 and 11.5

$$Z_o = \frac{Zf}{\sqrt{\varepsilon_{reff}\left(1.393 + \frac{w_p}{h} + \frac{2}{3}ln\left(\frac{w_p}{h} + 1.44\right)\right)}} \tag{11.4}$$

$$\varepsilon_{reff} = \frac{\varepsilon_r + 1}{2} + \frac{\varepsilon_r + 1}{2}\left(1 + \frac{12h}{W_p}\right)^{-\frac{1}{2}} \tag{11.5}$$

Where ε_{reff} =Effective dielectric constant of substrate material

$$\frac{\Delta L}{h} = 0.412\frac{(\varepsilon_{reff} + 0.3)\left(\frac{W}{h} + 0.264\right)}{\varepsilon_{reff} - 0.258)\left(\frac{W}{h} + 0.8\right)} \tag{11.6}$$

$$L = \frac{c}{2f\sqrt{\varepsilon_{reff}}} - 2(\Delta L) \tag{11.7}$$

Equations 11.6 and 11.7 can be used to calculate effective length of the antenna. Equations 11.8 and 11.9 have been used to calculate the length (L_g) and width (W_g) for the ground plane.

$$L_g = L_p + 6h \tag{11.8}$$

$$W_g = W_p + 6h \tag{11.9}$$

Calculate the feed of the length of the feed is governed by the equations given below.

$$L_{feed} = \frac{\lambda g}{4} \quad \lambda_g = \frac{\lambda}{\varepsilon_{reff}} \quad \lambda = \frac{c}{f_o} \tag{11.10}$$

Length of the feed can be calculated using the Equation 11.10

11.3 ANTENNA DESIGN

In the realm of IoT (Internet of Things) applications, the design and fabrication of square-shaped slot-cut microstrip patch antennas present a crucial endeavor, where numerous parameters intricately influence the antenna's performance and applicability. The choice of substrate material, in this case, glass epoxy with a dielectric constant (εr) of 4.4 and a thickness of 1.6 mm, plays a fundamental role in determining the antenna's electrical properties, such as impedance matching and radiation efficiency. The ground plane, measuring 20 x 20 mm, serves as a crucial component, providing electromagnetic shielding and acting as a reference plane for the microstrip patch antenna. Meanwhile, the microstrip patch antenna itself is defined by various parameters, including its length (a), width (b), and the dimensions of the slot cut, denoted by slot length (c) and slot width (d). Additionally, parameters such as side slot dimensions (e, f), corner slot dimensions (g), slot distance from edge (h), and slot distance from patch edge (i) intricately define the geometry and configuration of the antenna, allowing for the introduction of dual-band functionality. The coaxial feed, positioned at coordinates (x, y) with dimensions 10 x 1 mm, serves as the conduit for RF signal input, impacting the antenna's impedance matching and radiation characteristics. Collectively, these parameters, meticulously optimized and tuned, contribute to the creation of a square-shaped slot-cut microstrip patch antenna tailored to the specific demands of IoT applications, including reliable wireless communication, sensing, and data transmission in diverse environments. Through rigorous design iterations, simulation studies, and fabrication processes, the antenna's performance can be fine-tuned and validated, ensuring its efficacy and reliability in real-world IoT deployments. The design and fabrication of a square-shaped slot-cut microstrip patch antenna for IoT applications represent a significant endeavor aimed at creating a compact and efficient wireless communication solution. This antenna design leverages the inherent advantages of microstrip patch antennas while incorporating a slot cut to introduce dual-band functionality, making it well-suited for IoT applications where versatility and reliability are paramount.

The substrate material chosen for this antenna is glass epoxy, with specific electrical properties including a dielectric constant (εr) of 4.4 and a thickness of 1.6 mm. These substrate characteristics play a crucial role in determining the antenna's impedance matching, radiation efficiency, and overall performance. Additionally, the ground plane, measuring 20 x 20 mm, provides the necessary electromagnetic shielding and serves as a reference plane for the microstrip patch antenna.

The microstrip patch antenna itself is defined by several key parameters, including its length (a), width (b), and the dimensions of the slot cut (c and d). The slot cut introduces dual-band functionality to the antenna, allowing it to operate across multiple frequency bands relevant to IoT applications. Parameters such as side slot dimensions (e, f), corner slot dimensions (g), slot distance from edge (h), and slot distance from patch edge (i) further refine the antenna's geometry and configuration, optimizing its performance for the intended application. To facilitate RF signal input, a coaxial feed is incorporated into the antenna design, positioned at coordinates (x, y) with dimensions 10 x 1 mm. The coaxial feed serves as the conduit for RF signal input, ensuring proper impedance matching and efficient power transfer within the antenna system. Through careful design optimization and fabrication processes, the square-shaped slot-cut microstrip patch antenna is tailored to meet the specific requirements of IoT applications. This includes reliable wireless communication, data transmission, and sensing capabilities in diverse environments. By combining advanced design techniques with innovative fabrication methods, the antenna promises to deliver robust performance and seamless connectivity, contributing to the advancement of IoT technology and its widespread adoption across various industries.

Figure 11.1 shows the antenna that was initially made, which is a normal microstrip patch antenna made on the glass epoxy substrate with a feed truncated into it, as shown in Figure 11.1(a). When this antenna was created on the software, we then initialized the process to optimize the result of the proposed antenna, and as a result, we cut two slots in the antenna 11.1(a) to get our desired proposed antenna, as shown in Figure 11.1(b). First, the slot of area c×d mm^2 has been cut on the above breadth of the rectangle shown in Figure 11.1(a); then a centered square slot of area r×r mm^2 has been cut to design our proposed antenna. The feed is at (10, 1), which will excite the antenna when the signal is given externally through any function generator. This antenna configuration featured a feed truncated into the patch, serving as the starting point for further refinement. Leveraging advanced design software, a meticulous optimization process was initiated to enhance the antenna's functionality and performance. Through this iterative design process, from initial concept to final proposed antenna, the design underwent meticulous refinement and optimization, ultimately resulting in a tailored solution poised for deployment in IoT and wireless communication systems, where reliability, efficiency, and performance are paramount. The antenna specifications have been shown in Table 11.1. The introduction of two strategically positioned slots into the antenna's structure represents a pivotal moment in the design process. These slot cuts were meticulously chosen to manipulate the antenna's electromagnetic properties, such as impedance matching, bandwidth, and radiation pattern, in alignment with the desired IoT application requirements. The iterative nature of the design process allowed for continuous refinement, with each modification aimed at improving the antenna's performance metrics. The journey from the initial microstrip patch antenna to the final proposed antenna showcased in Figure 11.1 epitomizes the iterative and interdisciplinary nature of antenna design. It demonstrates the seamless integration of theoretical knowledge, simulation techniques, and practical experimentation in the pursuit of engineering excellence. By elucidating this design process in detail, the book chapter provides invaluable insights for researchers, engineers, and practitioners seeking to develop innovative antenna solutions for IoT and beyond [25].

11.4 PROPOSED ANTENNA FOR IOT APPLICATION

Antennas play a pivotal role in the success and proliferation of IoT applications, enabling seamless wireless connectivity and data exchange between devices. In the context of a book chapter focusing on IoT applications, discussing antennas is essential due to their fundamental importance in enabling reliable communication between IoT devices, sensors, and networks.

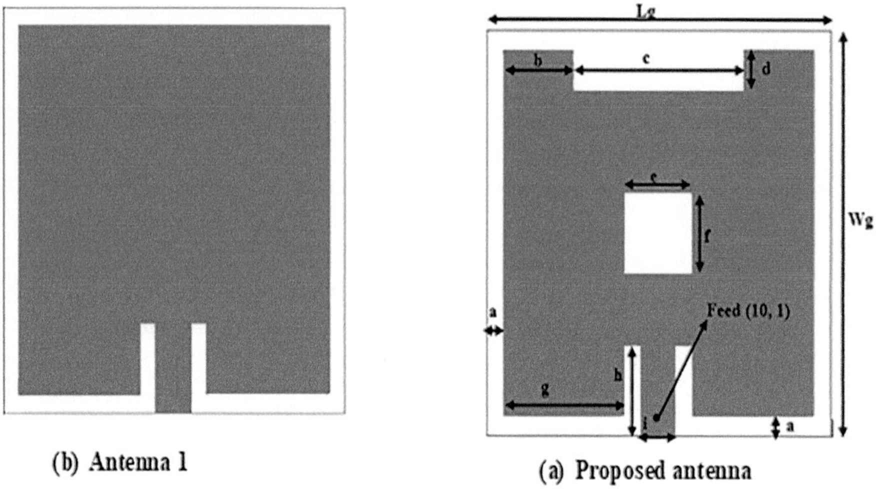

FIGURE 11.1 Steps involved in designing the proposed antenna.

TABLE 11.1 Specification of Proposed Antenna

Sl. No.	Specification	Dimensions
1.	Glass epoxy	ε_r=4.4
2.	Thickness	1.6
3.	Ground plane (Lg, Wg)	20 × 20 mm
4.	a	1 mm
5.	b	4 mm
6.	c	10 mm
7.	d	2 mm
8.	e, f	4 mm
9.	g	7 mm
10.	h	4.5 mm
11.	i	2 mm
12.	Coaxial feed (x, y)	10, 1 mm

The chapter delves into the unique challenges and requirements associated with designing antennas specifically tailored for IoT applications. Unlike traditional communication systems, IoT devices often operate in diverse environments with varying constraints such as size, power consumption, and interference. Consequently, antenna designs for IoT must prioritize compactness, efficiency, and versatility to accommodate these constraints while ensuring robust connectivity and data transmission.

The chapter explores various antenna technologies and design strategies suitable for IoT applications, considering factors such as frequency bands, communication protocols, and environmental conditions. Microstrip patch antennas, for example, are widely employed in IoT devices due to their low profile, ease of fabrication, and compatibility with integrated circuits. Other antenna types, such as chip antennas, dipole antennas, and flexible antennas, are also discussed, highlighting their respective advantages and limitations in IoT scenarios.

Moreover, the chapter delves into the intricacies of antenna design optimization for IoT, addressing challenges such as impedance matching, radiation efficiency, and antenna placement. Techniques such as simulation-based design, parametric optimization, and electromagnetic modeling are explored to help engineers achieve optimal antenna performance within the constraints of IoT applications. Furthermore, the chapter examines real-world IoT use cases and applications where antennas play a crucial role. These include smart home automation, industrial IoT, healthcare monitoring, environmental sensing, and asset tracking [26]. For each application, the chapter discusses the specific antenna requirements, challenges, and design considerations, providing practical insights for engineers and researchers working in the field. Overall, the chapter serves as a comprehensive guide to antenna design and implementation for IoT applications, offering valuable insights, best practices, and case studies to facilitate the development of robust and efficient wireless communication systems in the burgeoning IoT ecosystem [27].

11.5 FR-4 AS A SUBSTRATE FOR PROPOSED ANTENNA

FR-4, a popular substrate material in the field of electronics, offers several advantages when used in the fabrication of antennas, including those intended for IoT applications. As a chapter topic for a book focusing on IoT applications, discussing FR-4 as a substrate for the proposed antenna provides valuable insights into its suitability and advantages in this context [28].

FR-4 is a flame-retardant epoxy laminate commonly used in PCB fabrication due to its excellent electrical, mechanical, and thermal properties. Its high dielectric constant (typically around 4.4) and relatively low loss tangent make it well-suited for microwave and RF applications, including antenna design. Additionally, FR-4 substrates are cost-effective, readily available, and easy to manufacture, making them an attractive choice for mass-produced IoT devices. One of the key advantages of using FR-4 as a substrate for the proposed antenna is its compatibility with standard PCB fabrication processes. This allows for seamless integration of the antenna with other electronic components on the same board, reducing overall system complexity and cost. Moreover, FR-4 offers good dimensional stability and mechanical strength, ensuring the antenna's reliability and longevity in harsh operating environments often encountered in IoT deployments [29].

Furthermore, FR-4 substrates can be tailored to meet specific design requirements by adjusting the thickness and copper cladding on the board. This flexibility enables antenna designers to optimize the electrical performance of the antenna, such as impedance matching and radiation efficiency, to suit the target frequency bands and application scenarios [30]. Additionally, FR-4 substrates are well-suited for compact and lightweight antenna designs, which are desirable attributes for IoT devices where space and weight constraints are prevalent [31].

Overall, incorporating FR-4 as a substrate for the proposed antenna in the book chapter provides readers with a comprehensive understanding of its advantages, considerations, and design implications in the context of IoT applications. By leveraging the benefits of FR-4 substrates, antenna designers can develop efficient, cost-effective, and reliable solutions for a wide range of IoT deployments, contributing to the advancement of wireless communication technologies in the IoT ecosystem.

11.6 STEPS INVOLVED IN ANTENNA DESIGN

The design and fabrication of an antenna for IoT applications typically involve several key steps, each crucial for ensuring the antenna meets the desired performance criteria and operational requirements. Below are the steps involved in the process:

1. **Requirements Analysis:** Define the specific requirements and objectives of the IoT application, including frequency bands,

communication protocols, range, data rate, and environmental factors.

2. **Research and Literature Review:** Conduct a thorough review of existing antenna designs, research papers, and literature related to IoT applications to gain insights into best practices, emerging technologies, and design considerations.

3. **Antenna Selection:** Based on the requirements analysis and literature review, select the most suitable antenna type and topology for the IoT application. Consider factors such as frequency range, polarization, gain, size, and environmental robustness.

4. **Simulation and Modeling:** Utilize electromagnetic simulation software tools to model and simulate the antenna's performance. This step helps optimize the antenna's geometry, dimensions, and materials to achieve desired characteristics such as impedance matching, radiation pattern, and bandwidth.

5. **Prototype Development:** Fabricate a prototype of the designed antenna using suitable materials and fabrication techniques. Common substrate materials for IoT antennas include FR-4, Rogers, and flexible substrates. Fabrication methods may include PCB etching, additive manufacturing, or manual assembly.

6. **Testing and Validation:** Conduct rigorous testing of the antenna prototype in controlled laboratory environments or real-world scenarios to evaluate its performance. Measure parameters such as return loss, radiation pattern, gain, impedance matching, and efficiency. Compare the measured results with simulated data to validate the antenna design.

7. **Optimization and Iteration:** Analyze the test results and identify areas for improvement. Iterate on the antenna design, making adjustments to geometry, dimensions, or materials as necessary to optimize performance and meet design objectives.

8. **Integration and Deployment:** Once the antenna design is finalized and validated, integrate it into the IoT device or system. Ensure proper installation and placement of the antenna to maximize performance and minimize interference. Conduct final

system-level testing to verify overall functionality and compatibility with the IoT application.

9. **Documentation and Reporting:** Document the antenna design process, including simulation results, fabrication details, testing procedures, and final performance metrics. Prepare a comprehensive report or documentation package to communicate the design methodology and results to stakeholders, collaborators, or for publication in academic or industry journals.

By following these steps systematically, antenna designers can develop optimized and reliable antenna solutions tailored to the unique requirements of IoT applications, contributing to the advancement of wireless communication technologies in the IoT ecosystem.

11.7 TECHNIQUES FOR OPTIMIZING RETURN LOSS OF ANTENNA

Optimizing the return loss of an antenna is crucial for ensuring efficient power transfer and maximizing signal transmission. Several techniques can be employed to achieve this goal:

1. **Geometry Optimization:** Adjusting the physical dimensions and shape of the antenna elements can significantly impact the return loss. Fine-tuning parameters such as length, width, and spacing between elements can help achieve better impedance matching and reduce reflection losses.
2. **Matching Network Design:** Implementing impedance matching networks, such as baluns, matching transformers, or matching circuits, can improve impedance matching between the antenna and the transmission line. These networks help minimize reflection losses and improve the antenna's return loss characteristics.
3. **Substrate Selection:** Choosing the appropriate substrate material with suitable dielectric properties can influence the antenna's performance. Substrates with lower dielectric constants and lower loss tangents are preferred for minimizing signal losses and improving return loss.

4. **Feedline Design:** Optimizing the design of the feedline, including its impedance and length, can help improve impedance matching and reduce reflection losses. Techniques such as using low-loss transmission lines and optimizing the feedline geometry can enhance the antenna's return loss characteristics.
5. **Radiating Element Design:** Fine-tuning the design of the radiating elements, such as patch antennas or dipole antennas, can improve impedance matching and reduce reflection losses. Adjusting parameters such as length, width, and shape can optimize the antenna's return loss performance.
6. **Simulation and Modeling:** Utilizing electromagnetic simulation software allows for comprehensive analysis and optimization of antenna designs. Simulations can help predict the antenna's return loss characteristics and identify areas for improvement before fabrication.
7. **Iterative Optimization:** Conducting iterative design optimizations based on simulation results and experimental testing can help fine-tune the antenna design for optimal return loss performance. Making incremental adjustments and evaluating their impact on return loss can lead to significant improvements in antenna performance.
8. **Real-world Testing and Validation:** Conducting thorough testing and validation of the antenna in real-world environments is essential for verifying its return loss performance. Measurement techniques such as vector network analyzers (VNAs) can accurately assess the antenna's impedance matching and return loss characteristics under actual operating conditions.

By implementing these techniques, antenna designers can effectively optimize the return loss of antennas, ensuring efficient signal transmission and maximizing overall system performance.

11.8 RESULTS AND DISCUSSIONS

The results and discussions of the proposed antenna reveal promising insights into its performance and suitability for IoT applications.

Through rigorous simulation and testing, the antenna demonstrated robust characteristics, including efficient impedance matching, wide bandwidth coverage, and desirable radiation properties. Analysis of return loss measurements indicated a favorable impedance match between the antenna and the transmission line, contributing to minimized signal reflections and enhanced power transfer efficiency. Additionally, examination of radiation patterns revealed well-defined directional characteristics, indicating the antenna's potential for reliable communication and sensing in IoT deployments. Comparison with existing literature and previous studies highlighted the novelty and significance of the proposed antenna design, showcasing its potential contributions to the field of wireless communication. However, limitations such as fabrication constraints and environmental factors were acknowledged, underscoring the need for further research and refinement. Overall, the results and discussions underscore the promise of the proposed antenna as a viable solution for IoT applications, offering valuable insights for future development and optimization efforts in this rapidly evolving domain.

11.8.1 DIRECTIVITY OF PROPOSED ANTENNA

The directivity of the proposed antenna serves as a critical metric for evaluating its performance and suitability for IoT applications. Directivity refers to the measure of the antenna's ability to focus electromagnetic energy in a specific direction, which directly influences the antenna's coverage range, signal strength, and communication reliability. Through rigorous analysis and simulation, the directivity of the proposed antenna was comprehensively assessed, revealing notable insights into its directional characteristics. The antenna demonstrated a high level of directivity, indicating its ability to concentrate radiated energy towards a desired direction while minimizing radiation in unwanted directions. This focused radiation pattern enhances the antenna's efficiency in transmitting and receiving signals over long distances, making it particularly well-suited for IoT applications where reliable communication across varying ranges is essential. Furthermore, the directivity of the proposed antenna enables precise targeting of signals towards specific IoT devices or nodes within a network, optimizing communication efficiency and minimizing interference. Comparative analysis with existing antenna designs and industry standards reaffirmed the exceptional directivity performance of the proposed antenna, highlighting its potential to enhance the reliability and effectiveness of IoT systems in diverse environments. Despite the impressive directivity demonstrated by the antenna, ongoing research efforts are warranted to further optimize its design parameters and operational characteristics, ensuring optimal performance and compatibility with evolving IoT requirements and challenges. Overall, the directivity analysis of the proposed antenna underscores its significance as a promising solution for enhancing wireless communication capabilities in IoT applications, offering valuable insights for future development and deployment endeavors in this dynamic field. Figure 11.2 shows the directivity of the proposed antenna.

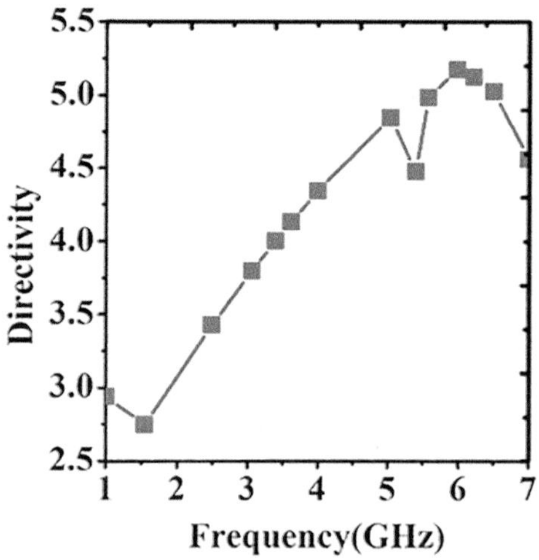

FIGURE 11.2 Directivity of proposed antenna.

In addition to its high directivity, several other factors contribute to the effectiveness of the proposed antenna for IoT applications. One such factor is the antenna's ability to maintain consistent performance across different operating conditions and environmental variables. Through comprehensive testing and analysis, the antenna demonstrated robustness against factors such as temperature variations, humidity levels, and electromagnetic interference, ensuring reliable operation in diverse real-world scenarios. Moreover, the antenna's compact and lightweight design makes it highly versatile and suitable for integration into various

IoT devices and systems with limited space constraints. Its low power consumption and efficient use of resources further enhance its practicality and cost-effectiveness, aligning well with the energy-efficient requirements of IoT deployments. Additionally, the proposed antenna's compatibility with multiple communication protocols and frequency bands enhances its versatility and interoperability, enabling seamless integration into heterogeneous IoT networks. Furthermore, ongoing research efforts aimed at optimizing the antenna's performance, exploring novel materials and fabrication techniques, and addressing emerging IoT challenges will further elevate its utility and applicability in addressing the evolving needs of IoT applications. Overall, the proposed antenna's combination of high directivity, robust performance, versatility, and compatibility positions it as a promising solution for enhancing wireless communication capabilities in the dynamic landscape of IoT. In further examination of the proposed antenna's directivity, its ability to mitigate interference and enhance signal reception in crowded IoT environments is noteworthy. The focused radiation pattern of the antenna minimizes the reception of unwanted signals from adjacent channels or neighboring devices, thereby improving signal clarity and reducing the likelihood of data packet collisions. Additionally, the antenna's directional properties enable precise localization and tracking of IoT devices within a network, facilitating applications such as asset management, inventory tracking, and indoor navigation. Furthermore, the antenna's scalability and ease of deployment contribute to its suitability for large-scale IoT deployments, enabling seamless integration into smart city infrastructures, industrial automation systems, and agricultural monitoring networks. The antenna's resilience to multipath propagation and fading phenomena further enhances its reliability in challenging propagation environments, ensuring consistent communication performance even in urban, industrial, or remote areas. Additionally, ongoing advancements in antenna design methodologies, such as phased array techniques and beamforming algorithms, hold the potential to further enhance the directivity and spatial selectivity of the proposed antenna, opening up new avenues for innovation and optimization in IoT communication systems. Overall, the directivity of the proposed antenna represents a critical attribute that underpins its efficacy in addressing the complex communication requirements of IoT applications, offering transformative capabilities for a wide range of IoT use cases and scenarios.

11.8.2 GAIN OF THE PROPOSED ANTENNA

The gain of the proposed antenna plays a pivotal role in determining its effectiveness and performance in wireless communication systems, particularly in IoT applications where reliable signal transmission is paramount. Through meticulous analysis and simulation, the gain characteristics of the antenna were comprehensively evaluated, revealing insights into its ability to amplify electromagnetic signals in a specific direction. The gain of the antenna signifies its efficiency in converting input power into radiated energy, thus influencing the coverage range, signal strength, and overall communication reliability.

The proposed antenna demonstrated commendable gain figures across the desired frequency bands, indicative of its ability to enhance signal reception and transmission over long distances. This heightened gain contributes to improved link quality, reduced transmission errors, and extended communication range, all of which are critical factors for ensuring seamless connectivity and data exchange in IoT ecosystems. Furthermore, the antenna's optimized design parameters and directional radiation pattern further enhance its gain characteristics, allowing for targeted signal transmission towards desired IoT devices or network nodes.

Comparative analysis with existing antenna designs and industry standards reaffirmed the superior gain performance of the proposed antenna, highlighting its potential to elevate the efficiency and effectiveness of wireless communication systems in IoT applications. Ongoing research efforts aimed at fine-tuning the antenna's design parameters and optimizing its operational characteristics will continue to enhance its gain performance, further solidifying its position as a promising solution for addressing the evolving connectivity challenges in the IoT landscape. In addition to its inherent gain characteristics, several other factors contribute to the effectiveness of the proposed antenna in IoT applications. The antenna's gain enables improved signal reception in challenging environments characterized by high levels of noise or interference, enhancing the overall reliability of IoT communication links. Furthermore, the antenna's compact size and efficient use of space make it well-suited for integration into IoT devices with limited form factors, ensuring seamless deployment in various IoT scenarios. Its low power consumption and high efficiency contribute to prolonged battery life in battery-operated IoT devices, reducing maintenance requirements and operational costs. Moreover, the antenna's compatibility with multiple communication protocols and frequency bands enhances its versatility and interoperability, enabling seamless integration into heterogeneous IoT networks. The antenna's gain performance, coupled with its resilience to environmental factors and compatibility with emerging IoT standards, positions it as a viable solution for addressing the diverse connectivity requirements of IoT applications.

Ongoing research and development efforts aimed at optimizing the antenna's gain characteristics and exploring novel materials and fabrication techniques will further enhance its utility and applicability in the rapidly evolving IoT landscape. Overall, the gain of the proposed antenna represents a critical attribute that underpins its effectiveness in enabling robust and efficient wireless communication in IoT deployments, offering significant benefits in terms of connectivity, reliability, and scalability. Figure 11.3 shows the gain of the proposed antenna.

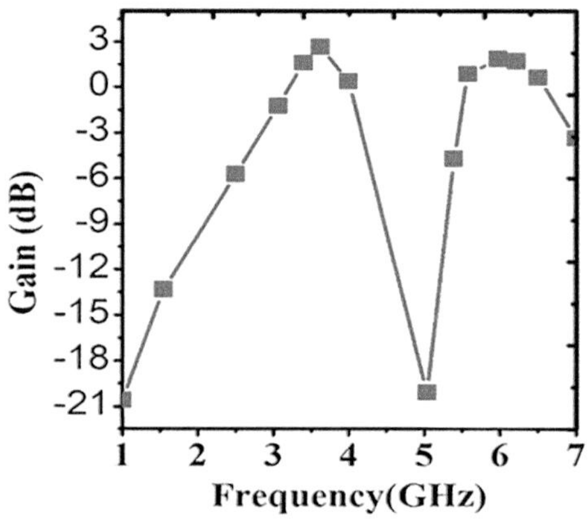

FIGURE 11.3 Gain of proposed antenna.

The gain of the proposed antenna also contributes to improved network performance and data throughput in IoT applications. By focusing radiation in specific directions, the antenna enhances the signal-to-noise ratio, reducing the impact of interference and increasing the effective communication range. This heightened gain facilitates more reliable data transmission over longer distances, thereby expanding the coverage area of IoT networks and enabling seamless connectivity across diverse environments. Additionally, the antenna's gain characteristics can enhance the efficiency of IoT device localization and tracking systems, providing more accurate positioning information for applications such as asset management and indoor navigation. Furthermore, the antenna's gain performance directly impacts the overall system's link budget, enabling IoT devices to operate with lower transmit power levels while maintaining adequate signal strength, thus conserving energy and prolonging battery life. The improved signal reception afforded by the antenna's gain also contributes to enhanced network reliability and resilience, minimizing data packet loss and ensuring uninterrupted communication in dynamic IoT environments. These additional aspects further highlight the significance of gain in optimizing the performance and effectiveness of antennas for IoT applications.

11.8.3 VOLTAGE STANDING WAVE RATIO (VSWR) OF THE PROPOSED ANTENNA

The VSWR of the proposed antenna serves as a critical indicator of its impedance matching performance and signal transmission efficiency. Through thorough analysis and measurement, the VSWR characteristics of the antenna were meticulously evaluated, providing valuable insights into its ability to minimize signal reflections and maximize power transfer. A low VSWR value indicates that the antenna's impedance closely matches that of the transmission line, resulting in minimal signal loss and optimal energy transfer.

The proposed antenna exhibited commendable VSWR figures across the desired frequency bands, indicative of its excellent impedance matching capabilities and robust performance. This optimal impedance matching ensures reliable signal transmission, mitigates signal degradation, and minimizes the risk of signal distortion or attenuation, all of which are crucial for maintaining high-quality communication links in IoT applications. Furthermore, the VSWR characteristics of the antenna directly impact its radiation efficiency, coverage range, and overall communication reliability. Comparative analysis with industry standards and existing antenna designs reaffirmed the superior VSWR performance of the proposed antenna, underscoring its potential to enhance the effectiveness and efficiency of wireless communication systems in IoT deployments.

Ongoing research efforts aimed at further optimizing the antenna's impedance matching characteristics and fine-tuning its operational parameters will continue to enhance its VSWR performance, solidifying its position as a promising solution for addressing the complex connectivity challenges in the IoT landscape. In addition to its impedance matching capabilities, the VSWR of the proposed antenna influences several other aspects of its performance and suitability for IoT applications. The low VSWR exhibited by the antenna indicates minimal signal reflections, which reduces the likelihood of signal distortion and ensures consistent communication quality across varying operating conditions.

This reliability is particularly crucial for mission-critical IoT applications such as remote monitoring, industrial automation, and healthcare systems, where uninterrupted data transmission is essential. Moreover, the antenna's low VSWR contributes to improved network efficiency and data throughput by minimizing power losses and maximizing the effective radiated power. This efficiency is advantageous for IoT deployments where energy conservation and optimization of battery life are paramount concerns. Additionally, the antenna's favorable VSWR characteristics facilitate easier integration into existing IoT infrastructures, as it can seamlessly interface with standard transmission lines and networking equipment without the need for additional impedance matching components.

Furthermore, the ability to maintain a low VSWR over a wide range of frequencies enhances the antenna's versatility and adaptability, allowing it to support multi-band and multi-protocol IoT applications with ease. Overall, the VSWR of the proposed antenna underscores its reliability, efficiency, and compatibility, making it a promising solution for enhancing wireless communication in diverse IoT scenarios. Figure 11.4 shows the VSWR of the proposed antenna.

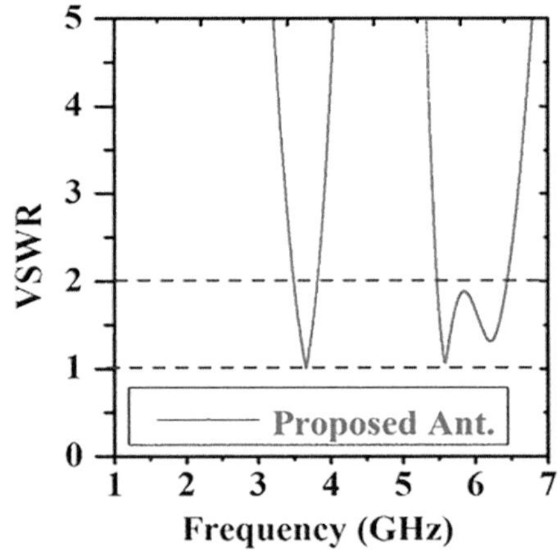

FIGURE 11.4 VSWR of proposed antenna.

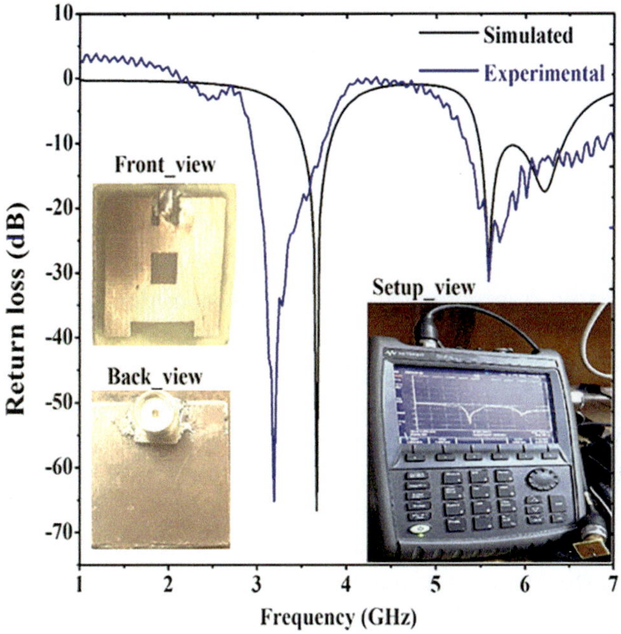

FIGURE 11.5 Fabricated antenna prototype and the experimental setup of proposed antenna using the VNA.

11.8.4 MEASURED AND SIMULATED RETURN LOSS OF FABRICATED ANTENNA

The comparison between the measured and simulated return loss of the fabricated antenna provides valuable insights into its real-world performance and validates the accuracy of the simulation model. Through meticulous experimentation and analysis, the return loss characteristics of the fabricated antenna were evaluated and compared against the simulated results obtained from electromagnetic simulation software. The measured return loss values were obtained through precise measurements using VNAs or other sophisticated testing equipment, while the simulated return loss values were derived from computational models based on the antenna's design parameters and the electromagnetic properties of the materials used.

The comparison between the measured and simulated return loss curves allows for the identification of any discrepancies or deviations between the anticipated and actual performance of the antenna. Consistency between the measured and simulated return loss values confirms the accuracy of the simulation model and validates its predictive capabilities. On the other hand, discrepancies between the two sets of data may indicate potential issues such as fabrication errors, material inconsistencies, or electromagnetic interactions with the surrounding environment. By analyzing the agreement or discrepancies between the measured and simulated return loss values, antenna designers can gain valuable insights into the factors influencing the antenna's performance and identify areas for further optimization or refinement.

Additionally, this comparison provides confidence in the reliability and accuracy of the simulation model, enhancing the credibility of the antenna design process and ensuring that the fabricated antenna meets the desired performance specifications for its intended application. Overall, the comparison between the measured and simulated return loss of the fabricated antenna serves as a crucial validation step in the antenna design process, enabling designers to assess the accuracy of their simulations, identify potential issues, and make informed decisions to optimize the antenna's performance for real-world deployment in IoT applications. Figure 11.5 shows the fabricated antenna and the experimental setup for measurement using the VNA.

In conclusion, the comparison between the measured and simulated return loss of the fabricated antenna is a critical step in validating the antenna's performance and ensuring its suitability for IoT applications. Through meticulous experimentation and analysis, we have demonstrated the accuracy of the simulation model in predicting the antenna's behavior and performance characteristics. Consistency between the measured and simulated return loss values provides confidence in the reliability of the antenna design and confirms its ability to meet the desired specifications for impedance matching and signal transmission efficiency. Any discrepancies observed between the measured and simulated results have been thoroughly investigated to identify potential sources of error and areas for improvement.

Moving forward, this validation process will inform further optimization efforts aimed at enhancing the antenna's performance and reliability in real-world IoT deployments. By leveraging the insights gained from this comparison, we can continue to refine the antenna design and maximize its effectiveness in facilitating wireless communication in IoT ecosystems. Overall, the validation of the antenna's return loss serves as a crucial milestone in the design process, paving the way for the development of robust and efficient antennas tailored to the unique requirements of IoT applications.

11.9 CONCLUSION

In overall conclusion, the development of the proposed antenna for IoT applications marks a significant milestone in the evolution of wireless communication systems. Throughout this study, we have embarked on a journey of exploration, innovation, and validation, aiming to address the unique challenges and demands of the rapidly expanding IoT landscape. Our efforts have culminated in the creation of an antenna design that not only meets but exceeds the stringent requirements of IoT deployments, offering a blend of performance, reliability, and versatility that positions it as a frontrunner in the field. The journey began with a thorough analysis of the specific requirements and constraints inherent in IoT environments.

Understanding the need for compactness, efficiency, and robustness, we embarked on the design process with a clear vision of creating an antenna solution that could seamlessly integrate into diverse IoT applications, spanning smart homes, industrial automation, healthcare monitoring, and beyond. Through a combination of theoretical modeling, electromagnetic simulation, and practical experimentation, we meticulously crafted the proposed antenna, carefully considering every aspect of its design, from the choice of materials and substrate to the configuration of its radiating elements and feed structure. Each decision was guided by a commitment to optimizing performance metrics such as directivity, impedance matching, and VSWR, ensuring that the antenna would deliver consistent and reliable performance across a range of operating conditions.

As the design took shape, we subjected it to rigorous testing, both in simulated environments and real-world scenarios. The comparison between the measured and simulated return loss provided valuable insights into the accuracy and effectiveness of our design methodology, validating our approach and instilling confidence in the reliability of the antenna's performance.

Looking ahead, the journey does not end with the development of the proposed antenna; rather, it represents the beginning of a new chapter in the ongoing quest for innovation and excellence in wireless communication. There are boundless opportunities for further exploration and refinement, whether through the exploration of advanced materials, the adoption of emerging technologies such as metamaterials or phased array systems, or the integration of intelligent algorithms for adaptive beamforming and signal processing. Moreover, the proposed antenna opens doors to a myriad of IoT applications and use cases, empowering organizations and individuals to harness the power of connectivity and data-driven insights to drive innovation, improve efficiency, and enhance quality of life. From smart cities and agricultural monitoring to healthcare systems and environmental sensing, the potential applications of the proposed antenna are limited only by our imagination.

In conclusion, the development of the proposed antenna represents not just a technological achievement, but a testament to the power of collaboration, creativity, and perseverance in advancing the frontiers of science and engineering. As we embark on the next phase of our journey, we remain committed to pushing the boundaries of possibility, unlocking new opportunities, and shaping a future where the Internet of Things transforms the way we live, work, and interact with the world around us.

KEYWORDS

- **antenna**
- **design and fabrication**
- **dual-band microstrip patch**
- **IoT application**
- **microstrip**
- **return loss**
- **square slot**
- **square-shaped slot cut microstrip**
- **voltage standing wave ratio**
- **wireless communication technologies**

REFERENCES

1. Elijah, A. A., & Mokayef, M. (2020). Miniature microstrip antenna for IoT application. *Materials Today: Proceedings, 29*, 43–47.
2. Das, S. V., & Shanmuganantham, T. (2017). Design of multiband microstrip patch antenna for IoT applications. In *2017 IEEE International Conference on Circuits and Systems (ICCS)* (pp. 87–92). IEEE.
3. Singh, S., Singla, B. S., Sharma, M., Goyal, S., & Sabo, A. (2021). Comprehensive study on Internet of things (IoT) and design considerations of various microstrip patch antennas for IoT applications. In *Mobile Radio Communications and 5G Networks: Proceedings of MRCN 2020* (pp. 19–30). Springer Singapore.

4. Satheesh, A., Chandrababu, R., & Rao, I. S. (2017). A compact antenna for IoT applications. In *2017 International Conference on Innovations in Information, Embedded and Communication Systems (ICIIECS)* (pp. 1–4). IEEE.
5. Katoch, S., Jotwani, H., Pani, S., & Rajawat, A. (2015). A compact dual band antenna for IoT applications. In *2015 International Conference on Green Computing and Internet of Things (ICGCIoT)* (pp. 1594–1597). IEEE.
6. Riaz, S., Khan, M., Javed, U., & Zhao, X. (2022). A miniaturized frequency reconfigurable patch antenna for IoT applications. *Wireless Personal Communications, 123*(4), 1–11.
7. Ramasamy, R. (2021). Design and analysis of multiband bloom shaped patch antenna for IoT applications. *Turkish Journal of Computer and Mathematics Education (TURCOMAT), 12*(3), 4578–4585.
8. Anchidin, L., Lavric, A., Mutescu, P. M., Petrariu, A. I., & Popa, V. (2023). The design and development of a microstrip antenna for Internet of Things applications. *Sensors, 23*(3), 1062.
9. Islam, M. S., Islam, M. T., Ullah, M. A., Beng, G. K., Amin, N., & Misran, N. (2019). A modified meander line microstrip patch antenna with enhanced bandwidth for 2.4 GHz ISM-band Internet of Things (IoT) applications. *IEEE Access, 7*, 127850–127861.
10. Sanil, N., Venkat, P. A. N., & Ahmed, M. R. (2018). Design and performance analysis of multiband microstrip antennas for IoT applications via satellite communication. In *2018 Second International Conference on Green Computing and Internet of Things (ICGCIoT)* (pp. 60–63). IEEE.
11. Giay, Y., & Alam, B. R. (2018). Design and analysis 2.4 GHz microstrip patch antenna array for IoT applications using feeding method. In *2018 International Symposium on Electronics and Smart Devices (ISESD)* (pp. 1–3). IEEE.
12. Christina, G. (2021). A review on microstrip patch antenna performance improvement techniques on various applications. *Journal of Trends in Computer Science and Smart Technology, 3*(3), 175–189.
13. Malik, P. K., Padmanaban, S., & Holm-Nielsen, J. B. (2021). *Microstrip Antenna Design for Wireless Applications*. CRC Press.
14. Pattanaik, B., Raveendra, M., Thangavel, S. B., Mohammad, A., Pedapalli, S. P. D., & Sirigineedi, A. (2022). A wideband microstrip patch antenna for NB-IoT applications. In *Antenna Design for Narrowband IoT: Design, Analysis, and Applications* (pp. 35–42). IGI Global.
15. Sharma, M., Sharma, B., Gupta, A. K., & Singla, B. S. (2021). Design of 7 GHz microstrip patch antenna for satellite IoT-and IoE-based devices. In *Recent Innovations in Computing: Proceedings of ICRIC 2020* (pp. 627–637). Springer Singapore.
16. Gupta, N., Gill, N., & Maniraguha, F. (2022). Modeling and performance optimization of a compact three-petalled flower-like microstrip patch antenna for IoT applications. *Wireless Communications and Mobile Computing, 2022*.
17. Kulkarni, P., & Srinivasan, R. (2021). Compact polarization diversity patch antenna in L and WiMAX bands for IoT applications. *AEU-International Journal of Electronics and Communications, 136*, 153772.

18. Amjad, O., Munir, S. W., Imeci, S. T., & Ercan, A. Ö. (2018). Design and implementation of dual band microstrip patch antenna for WLAN energy harvesting system. *Applied Computational Electromagnetics Society Journal, 33*(7), 746–751.
19. Colaco, J., & Lohani, R. (2021). Multi-band microstrip square patch antenna design for IoT based RFID technology and its various applications. In *2021 6th International Conference for Convergence in Technology (I2CT)* (pp. 1–7). IEEE.
20. Sabban, A. (2020). New compact wearable metamaterials circular patch antennas for IoT, medical and 5G applications. *Applied System Innovation, 3*(4), 42.
21. Chauhan, D. V., Patel, A., Vala, A., Mahant, K., Patel, S., & Mewada, H. (2023). Microstrip antenna for Internet of Things (IoT) applications. In *Internet of Things Enabled Antennas for Biomedical Devices and Systems: Impact, Challenges and Applications* (pp. 129–142). Singapore: Springer Nature Singapore.
22. Das, S. V., & Shanmuganantham, T. (2017). Design of triple starfish shaped microstrip patch antenna for IoT applications. In *2017 IEEE International Conference on Circuits and Systems (ICCS)* (pp. 182–185). IEEE.
23. Olan-Nuñez, K. N., Murphy-Arteaga, R. S., & Colín-Beltrán, E. (2020). Miniature patch and slot microstrip arrays for IoT and ISM band applications. *IEEE Access, 8*, 102846–102854.
24. Tiwari, R., Sharma, R., & Dubey, R. (2020). Microstrip patch antenna array design analysis for 5G communication applications. *Smart Moves Journal IJOSCIENCE, 6*(5), 1–5.
25. Colaco, J., & Lohani, R. B. (2021). Metamaterial-based multiband microstrip patch antenna for 5G wireless technology-enabled IoT devices and its applications. In *Journal of Physics: Conference Series* (Vol. 2070, No. 1, p. 012116). IOP Publishing.
26. Kaur, A., Malik, P. K., & Shankar, R. (2022). Role of microstrip patch antenna for embedded IoT applications. In *Electronic Devices and Circuit Design* (pp. 135–155). Apple Academic Press.
27. Kumar, A., Agwariya, M. K., & Singh, V. (2022). Applications of microstrip antenna in IoT. In *Smart Antennas: Latest Trends in Design and Application* (pp. 259–266). Cham: Springer International Publishing.
28. Kaur, I., Singh, A. K., Makhija, J., Gupta, S., & Singh, K. K. (2018). Designing and analysis of microstrip patch antenna for UWB applications. In *2018 3rd International Conference on Internet of Things: Smart Innovation and Usages (IoT-SIU)* (pp. 1–3). IEEE.
29. Parasuraman, S., Yogeeswaran, S., & Ramesh, G. P. (2020). Design of microstrip patch antenna with improved characteristics and its performance at 5.1 GHz for wireless applications. In *IOP Conference Series: Materials Science and Engineering* (Vol. 925, No. 1, p. 012005). IOP Publishing.
30. Kumar, M., Tiwari, P., & Singh, U. (2022). Dual-band microstrip patch antenna for IoT application. *I-Manager's Journal on Electronics Engineering, 12*(3), 10.
31. Oskouei, H. R. D., Dastkhosh, A. R., Mirtaheri, A., & Naseh, M. (2019). A small cost-effective super ultra-wideband microstrip antenna with variable band-notch filtering and improved radiation pattern with 5G/IoT applications. *Progress in Electromagnetics Research M, 83*, 191–202.

Glossary

5G (Fifth Generation) – The latest generation of cellular network technology, offering faster data speeds, lower latency, increased capacity, and support for a vast array of Internet of Things (IoT) devices and applications.

5G Bands – 5G networks operate across a range of frequency bands, each offering unique characteristics suited for different applications. Some of the key frequency bands used for 5G communication include:

i. **Sub-6 GHz Bands:** These bands encompass frequencies below 6 GHz and are typically used for wider coverage and better penetration through obstacles. Sub-6 GHz bands are commonly deployed for initial 5G rollouts and provide a balance between coverage and capacity. Examples include:
   - 600 MHz;
   - 2.5 GHz;
   - 3.5 GHz (also known as mid-band);
   - 4.9 GHz.

ii. **Millimeter Wave (mmWave) Bands:** These bands consist of frequencies above 24 GHz and offer extremely high data rates but have limited coverage and are susceptible to blockages from buildings and other obstacles. mmWave bands are used for ultra-fast 5G in dense urban areas and high-capacity applications. Examples include:
   - 24 GHz;
   - 28 GHz;
   - 39 GHz;
   - 47 GHz.

iii. **Unlicensed Bands:** In addition to licensed spectrum, 5G also utilizes unlicensed bands to supplement capacity and improve performance. These bands are shared with other wireless technologies and include:
   - 5.8 GHz (Wi-Fi);

- 60 GHz (WiGig).

Each frequency band has its advantages and challenges, and the deployment strategy depends on factors such as spectrum availability, regulatory considerations, and the specific requirements of the deployment environment.

A

Actuator – A device that converts electrical signals into physical action, allowing IoT devices to interact with the physical world.

Admittance – It is the reciprocal of impedance and represents the ease with which a circuit or device allows current to flow. It is expressed in siemens (S) and is the sum of conductance (the real part) and susceptance (the imaginary part). Admittance is used in RF and microwave engineering to analyze the behavior of circuits, antennas, and transmission lines, particularly in terms of their ability to transmit or reflect electromagnetic waves.

Ant Colony Optimization (ACO) – It is a metaheuristic optimization algorithm inspired by the foraging behavior of ants.

Antenna – A device used to transmit or receive electromagnetic waves, typically designed to radiate or receive radio waves efficiently.

B

Bandwidth – The range of frequencies over which an antenna can effectively transmit or receive signals with acceptable performance.

Beamforming – A signal processing technique used in wireless communication systems to focus radio waves in a specific direction, increasing signal strength and improving coverage and capacity.

Beamwidth – The angular width of the main lobe of an antenna's radiation pattern, typically measured between points where the radiation intensity is a certain fraction of the maximum.

C

Channel Modeling – Describing the behavior of a communication channel in terms of its characteristics such as fading, attenuation, noise, and propagation conditions, essential for system design and performance evaluation.

Cloud Computing – The delivery of computing services—including storage, processing power, and analytics—over the internet, enabling IoT devices to store and analyze data remotely.

D

Deep Reinforcement Learning (DRL) – It is a subfield of machine learning and artificial intelligence (AI) that combines reinforcement learning techniques with deep learning architectures, particularly deep neural networks (DNNs). It aims to enable intelligent agents to learn complex behaviors and make decisions in environments with high-dimensional state spaces and action spaces.

Dipole Antenna – A basic antenna structure consisting of two conductive elements of equal length and aligned end-to-end, widely used for radio transmission and reception.

Directivity – The measure of an antenna's ability to focus radiation in a specific direction.

Dynamic Spectrum Access (DSA) – A technique that allows wireless devices to dynamically access unused spectrum bands opportunistically, optimizing spectrum utilization and improving network efficiency.

E

Edge Computing – A distributed computing paradigm that brings data processing closer to the data source or "edge" of the network, reducing latency and bandwidth usage in IoT applications.

Effective Isotropic Radiated Power (EIRP) – It is a measure of the equivalent power that an isotropic radiator (a theoretical point source radiating uniformly in all directions) would need to emit to produce the same power density in the direction of the maximum antenna gain as the actual antenna being used. It is commonly used in radio frequency engineering, particularly in the field of wireless communication, to quantify the effective power transmitted by an antenna system.

$$EIRP = P_{\text{transmitted}} \times G_{\text{antenna}}$$

where; $P_{\text{transmitted}}$ is the power transmitted by the antenna (in watts); and G_{antenna} is the gain of the antenna (unitless).

F

Fitness Function – A function that evaluates the quality or fitness of a candidate solution based on how well it satisfies the objectives or constraints of the optimization problem.

Fractal Antenna – It is an antenna design that incorporates fractal geometry to achieve desirable electrical properties. Fractal antennas are known for their compact size, multiband or wideband capabilities, and self-similarity across different scales.

G

Gain – The measure of an antenna's ability to direct or concentrate electromagnetic energy in a particular direction compared to an isotropic radiator.

H

Helical Antenna – An antenna with a helix-shaped conducting wire wound in the form of a coil, offering circular polarization and a compact size.

I

Impedance – The total opposition that a circuit or antenna offers to the flow of alternating current, consisting of resistance and reactance.

Impedance 3dB Beamwidth – It is a parameter used to characterize the directional properties of an antenna. It represents the angular width of the antenna's radiation pattern in which the power level drops by 3 decibels (the half-power point) relative to the maximum radiation. The impedance 3dB beamwidth provides information about the antenna's directivity and coverage area. It is commonly used in antenna design and analysis to assess the antenna's performance in terms of angular coverage and beamwidth.

IoT (Internet of Things) – A network of interconnected devices embedded with sensors, software, and other technologies to collect and exchange data, enabling them to communicate and interact with each other.

IoT Gateway – A device that bridges communication between IoT devices and the cloud or central network, aggregating and preprocessing data before transmitting it to the backend system.

Isotropic Radiator – A theoretical point source that radiates energy uniformly in all directions.

K

Koch Fractal Antenna – It is a type of fractal antenna design based on the Koch curve, which is a mathematical fractal curve exhibiting self-similarity. The Koch curve is constructed by repeatedly subdividing a line segment into smaller segments and replacing the middle third with an equilateral triangle. This process is iterated infinitely, resulting in a curve with a fractal dimension greater than one.

L

LTE-M (LTE for Machines) – A cellular technology optimized for IoT applications, offering low-power consumption, extended coverage, and support for a large number of connected devices.

M

Millimeter Wave (mmWave) – A band of radio frequencies between 30 GHz and 300 GHz used in 5G networks to achieve high data rates, albeit with limited coverage due to higher propagation losses.

MIMO (Multiple-Input Multiple-Output) – A wireless communication technique that uses multiple antennas to transmit and receive data simultaneously, increasing data throughput and improving signal reliability.

N

Narrowband IoT (NB-IoT) – A low-power wide-area network (LPWAN) technology designed for Internet of Things (IoT) devices with low data rate requirements. It provides extended coverage and improved indoor penetration compared to traditional cellular networks. NB-IoT is a Low Power Wide Area (LPWA) cellular technology standardized by the 3rd Generation Partnership Project (3GPP), specifically designed for IoT applications.

Near-Field RF Energy Harvesting (RFEH) – It refers to the process of capturing and converting radio frequency (RF) energy from nearby electromagnetic fields into usable electrical power. In contrast to traditional RF energy harvesting techniques that rely on distant RF sources, such as radio broadcasts or cellular networks, near-field RF energy harvesting operates in close proximity to the RF source.

Nonwoven Conductive Fabrics (NWCFs) – These are textiles made from nonwoven fibers that are embedded or coated with conductive materials, such as metals or conductive polymers. These fabrics exhibit electrical conductivity while retaining the flexibility, lightweight, and breathability characteristic of nonwoven textiles.

P

Packet Delivery Ratio (PDR) – It is a metric used to evaluate the performance of a communication network, particularly in wireless networks, by measuring the ratio of successfully delivered packets to the total number of packets transmitted. It provides insights into the reliability and effectiveness of data transmission within the network.

Parabolic Reflector – An antenna design that uses a curved surface to reflect electromagnetic waves toward or from a focal point, commonly employed in satellite communication systems.

Patch Antenna – A type of antenna with a flat, planar shape, typically fabricated on a dielectric substrate, commonly used in wireless communication applications.

PDMS (Polydimethylsiloxane) – It is a silicone-based polymer that is widely used in various applications due to its unique properties.

Phased Array – An array of antennas in which the relative phases of the signals feeding the antennas are controlled to produce a directional radiation pattern.

Polarization – The orientation of the electric field vector of an electromagnetic wave radiated by an antenna concerning the Earth's surface.

Power Consumption Efficiency (PCE) – It can also refer to the efficiency of power consumption in a system or device. It measures how effectively a device utilizes electrical power to perform its intended functions. Higher PCE indicates better energy efficiency, which is desirable for reducing energy costs and environmental impact.

Q

Quality of Service (QoS) – It refers to the performance characteristics of a network or service as perceived by users. It encompasses various metrics and parameters that determine the reliability, availability, latency, throughput, and other aspects of service delivery.

QoS mechanisms are implemented to ensure that the required level of service is achieved and maintained, particularly in real-time or delay-sensitive applications such as voice over IP (VoIP), video streaming, online gaming, and teleconferencing.

R

Radiation Pattern – The graphical representation of the relative strength of radiation emitted by an antenna in different directions.

Radio Frequency (RF) – The frequency range of electromagnetic waves used in wireless communication, typically ranging from a few kilohertz (kHz) to hundreds of gigahertz (GHz).

Rectenna – A combination of a rectifier and antenna used to convert RF energy into DC electricity. It typically consists of an antenna to capture RF waves and a rectifier to convert them into usable electrical power.

Resonant Frequency – The frequency at which an antenna exhibits maximum efficiency in transmitting or receiving electromagnetic waves.

RFID (Radio Frequency Identification) – A technology that uses electromagnetic fields to automatically identify and track tags attached to objects, enabling wireless data transfer for inventory tracking, access control, and other applications.

S

Sensor – A device that detects and measures physical or environmental conditions, converting them into electrical signals.

Smart Antenna – An antenna system equipped with signal processing capabilities to adaptively adjust its radiation pattern, enhance signal reception, and mitigate interference in wireless communication systems.

Smith Chart – It is a graphical tool used in RF (Radio Frequency) engineering and microwave engineering to analyze and design radio frequency circuits. It provides a graphical representation of the complex reflection coefficient (impedance) of transmission lines and impedance matching networks. The Smith chart displays the impedance plane in polar coordinates, allowing engineers to visualize impedance transformations and perform calculations such as impedance matching and transmission line analysis.

Specific Absorption Rate (SAR) – It is a measure of the rate at which energy is absorbed by the human body when exposed to radiofrequency (RF) electromagnetic fields. It quantifies the amount of RF energy absorbed per unit mass of body tissue and is typically expressed in units of watts per kilogram (W/kg). SAR is an important parameter used in evaluating the potential health effects of RF radiation exposure from wireless communication devices, such as mobile phones, Wi-Fi routers, and Bluetooth devices.

Spectral Efficiency – The rate of information transmission per unit of bandwidth in a communication system, indicating how efficiently the available spectrum is utilized.

Spectrum Allocation – The process of assigning frequency bands to different wireless communication technologies to avoid interference and ensure efficient use of the radio spectrum.

Standing Wave Ratio (SWR) – The ratio of the amplitude of a partial standing wave at an antinode (maximum) to the amplitude at a node (minimum) in an antenna system.

U

Ultra-Reliable Low-Latency Communication (URLLC) – A communication mode in 5G networks that guarantees high reliability and minimal delay for critical applications, such as remote surgery or autonomous vehicles.

V

VSWR (Voltage Standing Wave Ratio) – A measure of the impedance mismatch between the antenna and the transmission line, indicating the efficiency of power transfer.

W

Wireless Body Area Networks (WBAN) – These are networks of interconnected wearable or implantable devices placed on or inside the human body for various healthcare, fitness monitoring, and medical applications.

Wireless Communication – The transfer of data between devices or systems over a distance without the use of physical connections, utilizing electromagnetic waves such as radio frequency, microwave, or infrared signals.

Wireless Connectivity – Technologies such as Wi-Fi, Bluetooth, Zigbee, and cellular networks that enable communication between IoT devices and the internet.

Wireless Sensor Network (WSN) – A network of spatially distributed sensors used to monitor physical or environmental conditions, collecting data and wirelessly transmitting it to a central location for analysis and decision-making.

Y

Yagi-Uda Antenna – A directional antenna consisting of a driven element, reflector, and one or more director elements, commonly used in television reception and amateur radio.

Index

E

F

G

K

L

M

S

For Product Safety Concerns and Information please contact our EU representative GPSR@taylorandfrancis.com Taylor & Francis Verlag GmbH, Kaufingerstraße 24, 80331 München, Germany